WEST GEORGIA TECH LIB
FORT DRIVE
LAGRANGE, GA 30240

.U44 industrial polymers /
5   Henri Ulrich.

WEST GEORGIA TECH LIBRARY
FORT DRIVE
LAGRANGE, GA 30240

Ulrich

Introduction to
Industrial Polymers

SPE Books from Hanser Publishers

Bernhardt, Computer Aided Engineering for Injection Molding
Brostow/Corneliussen, Failure of Plastics
Charrier, Polymeric Materials and Processing – Plastics, Elastomers and Composites
Ehrig, Plastics Recycling
Gordon, Total Quality Process Control for Injection Molding
Gruenwald, Plastics: How Structure Determines Properties
Macosko, Fundamentals of Reaction Injection Molding
Manzione, Applications of Computer Aided Engineering in Injection Molding
Matsuoka, Relaxation Phenomena in Polymers
Menges/Mohren, How to Make Injection Molds
O'Brien, Applications of Computer Modeling for Extrusion and Other Continuous Polymer Processes
Michaeli, Extrusion Dies for Plastics and Rubber
Rauwendaal, Polymer Extrusion
Saechtling, International Plastics Handbook for the Technologist, Engineer and User
Stoeckhert, Mold-Making Handbook for the Plastics Engineer
Throne, Thermoforming
Tucker, Fundamentals of Computer Modeling for Polymer Processing
Ulrich, Introduction to Industrial Polymers
Wright, Molded Thermosets: A Handbook for Plastics Engineers, Molders and Designers

Henri Ulrich

Introduction to Industrial Polymers

Second Edition

Hanser Publishers, Munich Vienna New York Barcelona

Dr. Henri Ulrich
Dow Chemical USA, North Haven Laboratories
North Haven, CT

The use of general descriptive names, trademarks, etc. in this publication, even if the former are not especially identified, is not to be taken as a sign that such names, as understood by the Trade Marks and Merchandise Marks Act, may accordingly be used freely by anyone.

While the advice and information in this book are believed to be true and accurate at the date of going to press, neither the authors nor the editors nor the publisher can accept any legal responsibility for any errors or omissions that may be made. The publisher makes no warranty, express or implied, with respect to the material contained herein.

Die Deutsche Bibliothek – CIP-Einheitsaufnahme

Ulrich, Henri:
Introduction to industrial polymers / Henri Ulrich. – 2. ed. –
Munich ; Vienna ; New York ; Barcelona : Hanser, 1993
 (SPE books)
 ISBN 3-446-17119-3

All rights reserved. No part of this book may be reproduced or transmitted in any form or by any means, electronic or mechanical, including photocopying or by any information storage and retrieval system, without permission in writing from the publisher.

Copyright © Carl Hanser Verlag, Munich, Vienna, New York, Barcelona 1993
Printed in Germany

Foreword

The Society of Plastics Engineers is pleased to sponsor and endorse this second edition of "Introduction to Industrial Polymers". The original edition, published ten years ago, was the first Hanser volume which SPE sponsored. As hoped in the preface of that book, the event marked the beginning of an outstanding relationship between the two organizations.

This second edition once again treats the commercial significance and utilization of polymers from an industrial, rather than an academic viewpoint. The revision includes extensive material on a variety of subjects which were barely emerging ten years ago including recycling, alloys and blends, liquid crystal polymers and biopolymers as well as data on several new commercial polyolefins.

SPE, through its Technical Volumes Committee, has long sponsored books on various aspects of plastics and polymers. Its involvement has ranged from identification of needed volumes to recruitment of authors. An ever-present ingredient, however, is review of the final manuscript to insure accuracy of the technical content.

This technical competence pervades all SPE activities, not only in publication of books but also in other activities such as technical conferences and educational programs. In addition, the Society publishes periodicals – *Plastics Engineering, Polymer Engineering and Science, Polymer Processing and Rheology, Journal of Vinyl Technology and Polymer Composites* – as well as conference proceedings and other selected publications, all of which are subject to the same rigorous technical review procedure.

The resource of some 37,000 practicing plastics engineers has made SPE the largest organization of its type worldwide. Further information is available from the Society at 14 Fairfield Drive, Brookfield, Connecticut 06804, U.S.A.

Robert D. Forger
Executive Director
Society of Plastics Engineers

Technical Volumes Committee
Raymond J. Ehrig, Chairperson
Aristech Chemical Corporation

Preface

In the early 1980s I taught a mini-course on industrial polymers at Wesleyan University in Middletown, Connecticut. My attempts to find a suitable textbook met with complete failure. Most textbooks in polymer science deal with principles and are slanted toward material science rather than chemistry. I felt that it was essential to indicate to the students the polymers that are used extensively in today's world, their significance and social impact, and future trends in the development of useful macromolecules.

Polymer science is still a young discipline. Hermann Staudinger, the first Nobel Prize winner in this field, did his pioneering work on the "existence" of macromolecules in the 1920s. Nevertheless, we are now in the plastic age. Plastics production, on a volume basis, surpassed steel production in the United States in 1979. The annual global plastics consumption today is in excess of 60 million metric tons. Recent years have seen the emergence of functional polymers used in the design of new drug delivery systems and in the synthesis of biopolymers, reinforced plastics are used increasingly in the production of modern aircraft, and polymeric membranes are used in the desalination of seawater and in the separation of industrial gases. The development of plastic compact discs has revolutionized the audio industry. Reaction polymers, such as polyurethanes, are used for the insulation of the main fuel tank of the space shuttle. Liquid crystal copolymers, introduced in the 1980s, provide self-reinforcing thermoplastics with properties superior to glass-reinforced plastics. New polymer processing technologies provide challenges to develop new families of liquid monomers to be used in reaction molding of new materials of construction.

This text ist divided into four parts: Part I provides an overview of the industry, and Parts II through IV deal with addition polymers, condensation polymers, and special polymers. Classification of the polymers according to polymerization technique and monomer composition provides an orderly approach, whereas classification based on polymer applications – that is, plastics, rubber materials (elastomers), fibers and coatings – would result in too much overlap. For example, polyamides are used mainly as synthetic fibers, but they have also found uses as engineering thermoplastics and as powder coatings. Polypropylene is not only used as a thermoplastic molding compound; it has also found applications as a fiber-forming material.

The vinyl compounds are all based on the homopolymerization of an olefin monomer. To differentiate them, the various types of olefin homopolymer are classified by the substituents attached to one carbon of the double bond. If the substituent ist hydrogen (as in polyethylene) or an alkyl or aryl group, the products are listed under polyolefins. If the substituent is a nitrile, a carboxylic acid, or a carboxylic acid ester, we are dealing with derivatives of acrylic acid, and the derived polymers are listed under acrylics. A third group of olefin homopolymers with substituents bonded to the double bond through an oxygen or nitrogen atom are described under polyvinyl compounds.

The use of trade names has been avoided whenever possible. However, many trade names have become household words (Nylon, Formica, Spandex, etc.), and they are used to identify the more common products.

The plastic materials used today are seldom true homopolymers. For example, polyethylenes are often copolymers, containing 1 to 10 % of another olefin monomer. Also numerous ethylene copolymers, containing higher amounts of a comonomer, are produced to tailor their properties to the needs of the intended applications. In additrion, alloys and blends of the major thermoplastic polymers are produced extensively today to create numerous new multipolymers. For example, thermoplastic polycarbonates undergo stress cracking upon exposure to gasoline. This phenomenon can be greatly reduced by alloying polycarbonate with PBT or other polyesters. These alloys are now being used to mold automotive parts. Elastomers can also be added as internal plasticizers to improve the impact strength of more brittle polymers.

The marketplace will experience increasingly cannibalistic competition among the major thermoplastics: growth at the expense of each other, rather than replacement of traditional materials. Recycling of plastics will slow the growth of plastic consumption because a good portion of the major thermoplastic materials (PET, PE, PP, PS) will be recycled. The availability of fossil-based raw materials will not be a determining factor in plastic growth. In spite of many prophets of doom, it is clear that plastics will continue to serve mankind in the many applications described in this book.

Guilford, Connecticut *Henri Ulrich*

Contents

Part I: General Introduction

1 Historical Overview of Industrial Polymers 13
2 Basic Polymer Principles 17
3 Raw Materials for the Polymer Industry 22
4 Production of Industrial Polymers 27
5 Processing of Industrial Polymers 30
6 Recycling of Industrial Polymers 35

Part II: Addition Polymers

7 Polyolefins 41
   Introduction 41
   Polyethylene (PE) 50
   Polyvinyl fluoride (PVF) 54
   Polyvinyl chloride (PVC) 55
   Polyvinylidene fluoride (PVDF) 56
   Polyvinylidene chloride (VDC) 57
   Polychlorotrifluoroethylene (PCTFE) 57
   Polytetrafluoroethylene (PTFE) 58
   Polypropylene (PP) 59
   Poly(1-butene) 60
   Poly(4-methylpentene) 60
   Polystyrene (PS) 61
   Polyvinyl pyridine 62
   Polybutadiene (Butadiene Rubber, BR) 63
   Polyisoprene 65
   Polychloroprene 66

8 Olefin Copolymers 68
   Introduction 68
   Styrene-Acrylonitrile Copolymers (SAN) 69
   Acrylonitrile-Butadiene-Styrene Terpolymers (ABS) 70
   Ethylene-Methacrylic Acid Copolymers (Ionomers) 71
   Styrene-Butadiene Rubber (SBR) 73
   Styrene-Butadiene Block Copolymers 74
   Nitrile Rubber (NBR) 76
   Ethylene-Propylene Elastomers 76
   Butyl Rubber 77
   Tetrafluoroethylene Copolymers 78
   Fluoroelastomers 79

9 Alloys and Blends 80
   Introduction 80

Polyolefin Alloys and Blends	81
Polyvinyl chloride Alloys and Blends	83
ABS and SAN Alloys and Blends	84
Nylon Alloys and Blends	85
Polycarbonate Alloys and Blends	85
Polyester Alloys and Blends	86
Polyphenylene Oxide Alloys and Blends	86
Fluoropolymer Alloys and Blends	87
10 Acrylics	88
Introduction	88
Acrylic Fibers	89
Acrylic Adhesives	90
Polyacrylates	91
Polymethyl methacrylate (PMMA)	93
Polyacrylamide	94
11 Polyvinyl Compounds	95
Introduction	95
Polyvinyl Acetate (PVA)	96
Polyvinyl Alcohol (PVAL)	97
Polyvinyl Butyral (PVB) and Polyvinyl Formal (PVF)	98
Polyvinyl Ethers	98
Polyvinylpyrrolidone (PVP)	99
Polyvinylcarbazole	99
12 Polyurethanes	100
Introduction	100
Flexible Polyurethane Foam	103
Rigid Polyurethane Foam	103
Polyurethane Elastomers	104
Polyurethane Coatings, Adhesives, and Sealants	106
13 Ether Polymers	109
Introduction	109
Polyacetal	109
Polyethylene Glycol (PEG) and Polypropylene Glycole (PPG)	110
Epoxy Resins	112
Polyphenylene Oxide (PPO)	116

Part III: Condensation Polymers

14 Polyesters	119
Introduction	119
Polyethylene terephthalate (PET)	122
Polybutylene terephthalate (PBT)	123
Polydihydroxymethylcyclohexyl terephthalate	124
Cellulose Esters	124
Unsaturated Polyesters	126

Aromatic Polyesters 127
Polycarbonate (PC) 128

15 Polyamides 131
   Introduction 131
   Aliphatic Polyamides 131
   Aliphatic-Aromatic Polyamides 135
   Aromatic Polyamides 135
   Polyamide Imides 138
   Polyimides 139

16 Formaldehyde Resins 141
   Introduction 141
   Phenol-Formaldehyde Resins (PF) 141
   Urea-Formaldehyde Resins (UF) 143
   Melamine-Formaldehyde Resins (MF) 144

Part IV: Special Polymers

17 Heat-Resistant Polymers 149
   Introduction 149
   Arylketones 151
   Polyphenylene Sulfide (PPS) 152
   Polysulfones 153
   Polybenzimidazole (PBI) 155
   Liquid Crystal Polymers (LCPs) 156

18 Silicones and Other Inorganic Polymers 157
   Introduction 157
   Silicones 158
   Polyphosphazenes (PNF) 160
   Polycarborane-siloxanes 160
   Polythiazyl 161

19 Functional Polymers 162
   Introduction 162
   Photoconductive Polymers 162
   Electroconductive Polymers 163
   Piezoelectric Polymers 165
   Light-Sensitive Polymers 165
   Hollow-Fiber Membranes 167
   Ion-Exchange Resins 167
   Polymeric Reagents 169
   Biopolymers 171

Appendix

Guide to Further Reading 176
Major World Producers of High Volume Industrial Polymers 179
Commonly Used Abbrevations for Industrial Polymers 184
Index 186

Part I
General Introduction

WEST GEORGIA TECH LIBRARY
FORT DRIVE
LAGRANGE, GA 30240

1 Historical Overview of Industrial Polymers

On a volume basis, U. S. plastics production in 1979 surpassed U. S. steel production, heralding the arrival of the plastics age in the United States. Plastics are still evolving, still changing and improving as the 21st century approaches. Their potential impact on society appears to be greater than ever. We have all seen their increasing use in everyday life: in apparel, home furnishings, housewares, contruction materials, tanks and pipes, insulation, storage, transportation, packaging, medical uses, recreational activities, communications, electronics, aerospace, and many other applications.

Natural polymeric products have been used widely throughout the ages, but synthetic polymers are of fairly recent vintage. Before today's sophistication in plastics could be achieved, the chemists of the last century had first to develop an understanding of natural polymers. The first hypothesis of the existence of macromolecules was advanced in 1877 when Kekulé proposed that natural organic substances consist of very long chains of molecules from which they derive their special properties. Interestingly, in 1868 John Wesley Hyatt had already manufactured the first synthetic plastic in the United States, producing celluloid by treating cotton whith nitric acid and camphor. This development was prompted by a shortage of ivory, which led Hyatt to seek a synthetic material for the production of billiard balls.

Celluloid later served as the base for the first films used in still photography and in the early motion pictures. In 1893 Emil Fischer proposed a structure for natural cellulose. He considered cellulose to contain a chain of glucose units. He also postulated that peptides are long-chain amino acids. Today's peptide synthesizers use polymeric substrates (Merrifield resins) to synthesize polypeptides consisting of more than a hundred amino acids in the chain. In 1909 another American scientist, Leo Hendrik Baekeland, invented phenol-formaldehyde resins. These materials, which became known as Bakelite, were used widely in the manufacture of molded parts, including the first mass-produced telephones.

The step from the idea of macromolecules to the reality of producing them at will was still not made, because both Hyatt and Baekeland invented their plastics by trial and error. It needed the genius of Hermann Staudinger, who in 1924 proposed linear structures for polystyrene and natural rubber. Later Staudinger received the Nobel Prize in chemistry for his pioneering work in macromolecular chemistry. After the recognition of the fact that macromolecules really are linear polymers, it did not take long for other materials to emerge. In 1927 cellulose acetate and polyvinyl chloride (PVC) were developed, and 1929 saw the production of urea-formaldehyde resins. In 1930 the U. S. production of the new industrial polymers was only 23,000 metric tons, mostly phenolics and celluloid.

Again it needed a scientist with a broad vision to provide a new concept for the development of synthetic polymers. This man was Wallace H. Carothers, who pioneered linear condensation polymers, such as polyesters and polyamides. Carothers joined Du Pont's central research laboratory in Wilmington, Delaware, in 1928, coming from Harvard University. He chose as his research project to study long-chain molecules made from difunctional monomers. One member of his group, Paul Flory, received the Nobel Prize in chemistry in 1974. The basic research efforts of the Carothers team led to the development of neoprene, polyesters, and polyamides. Du Pont's nylon production started in 1938, and nylon-6 (perlon) production by I. G. Farben began in 1939. Also in Germany, P. Schlack had developed the caprolactam route to nylon, which was the first example of a ring-opening polymerization.

The years prior to the entry of the United States into World War II saw the rapid development of many important plastics, such as acrylic polymers and polyvinyl acetate in 1936, polystyrene in 1938, melamine formaldehyde resins (formica) in 1939, and polyester and polyethylene in 1941.

The amazing scope of wartime applications led to the development of scratch-resistant, heat-stable, weatherproof, transparent plastics for aircraft, buildings, and containers; lightweight, waterproof, nonflammable, noncreasing synthetic fibers for garments, carpets, upholstery, and draperies; cellular plastic insulation and sandwich construction used in lightweight structural materials for walls, ceilings, and furniture; resins with improved heat resistance and electrical insulation properties in detection devices for aircraft, ships, and trains; plastic sheeting and fiber-reinforced materials fabricated into large, light, tough structures for cars and appliances; and elastomeric crosslinked copolymers to provide superior performance in wire insulation, weather-stripping, rainwear, hospital accessories, chemical-resistant coatings, and structural adhesives.

After the war, the development of new polymeric materials accelerated at an even faster pace. In 1947 epoxies were developed (experimental epoxy resins were already used in England during the war in the manufacture of De Havilland Hornet twin-engine fighters), and in 1948 acrylonitrile-butadiene-styrene terpolymers (ABS) were introduced. The polyurethanes, which were produced in Germany already in the 1930s, were rapidly developed in the United States as German technology became available after the war. By 1949 the annual production of industrial polymers in the United States reached 570,000 tons.

Again the input of a brilliant scientist stimulated polymer science, and this person was Karl Ziegler at the Max Planck Institut in Mühlheim, Ruhr, Germany. Ziegler was involved in synthetic organometallic chemistry. Exposure of ethylene to some aluminum alkyl compounds led to very rapid polymerization of the ethylene. Fifty years earlier an investigator would have discarded such an intractable material, but the time was right for macromolecules, and Ziegler recognized the importance of his findings. He and Giulio Natta extended the work to other olefins, and both chemists received the Nobel Prize in 1963 for their discovery of stereospecific polymerization.

The 1950s saw the emergence of linear polyethylene, polypropylene, polyacetal, polyethylene terephthalate, polycarbonate, and a host of new copolymers. The U. S. plastics production in 1959 reached 2,730,000 tons. The next decade saw the development of an incredible number of processing technologies accompanying the surge of plastics to a significant position as materials of construction. Significant new plastics launched in the 1970s include polyphenylene sulfide and polybutylene terephthalate. Also, more heat-resistant polymers, such as polyimides, polyarylketones, polyarylates, polyarylsulfones, polyarylamides, and polybenzimidazoles were introduced. In the 1980s liquid crystal copolymers emerged. These self-reinforcing polymers have superior properties over glass-reinforced composites. The highlights of the short development phase of industrial polymers are listed in Table 1.

Table 1 Historical development of Industrial Polymers*

Date	Event
1868	J. W. Hyatt invented cellulose nitrate and the injection-molding machine.
1877	A. Kekulé postulated that natural products are long-chain molecules.
1893	E. Fischer verified that the structure of cellulose is indeed a macromolecule.
1909	L. H. Baekeland invented PF resins (Bakelite).
1924	H. Staudinger proposed the linear chain structure for polystyrene.
1927	Introduction of cellulose acetate and PVC.
1928	O. Röhm commercialized polymethyl methacrylate (PMMA).
1929	Introduction of UF resins.
1930	First production of polystyrene (PS).
1935	W. H. Carothers first synthesized nylon-6,6.
1936	Introduction of PAN, SAN, and polyvinyl acetate.
1937	O. Bayer invented polyurethanes.
1938	P. Schlack invented nylon-6 and epoxy resins. R. J. Plunkett discoverd the polymerization of tetraflouroethylene. High-pressure polymerization of ethylene was introduced.
1940	G. E. Rochow invented the direct process for the manufacture of chlorosilanes, the raw materials for silicone resins.
1941	J. R. Whinfield and J. T. Dickinson invented polyethylene terephthalate (PET).
1942	Commercial introduction of PAN fibers (Orlon®).
1948	Introduction of ABS.
1952	K. Ziegler developed catalysts for the low-pressure polymerization of ethylene.
1953	G. Natta used Ziegler cataysts to synthesize stereoregular polypropylene (PP). H. Schnell invented polycarbonate.
1956	A. S. Hay discovered polyphenylene oxide (PPO).
1958	Commercial introduction of polyacetal.

* A list of abbreviations for industrial polymers is given at the end of the book.

The consumption of plastics in the automotive industry has surpassed 300 pounds per car, and by 1978 the annual production of industrial polymers in the United States exceeded 19 million tons.

Social and economic challenges have added a new dimension to macromolecules. The need for recycling of plastics was clearly recognized in the 1980s. All major manufacturers of plastic materials are involved in this undertaking, and recycling will have an impact on the growth of plastics. The petrochemical feedstock for plastics will be available well into the next century. Only 1.5 % of the oil and natural gas consumed annually in the United States is used in the production of plastics. However, future oil price escalation may have an even more significant impact on the growth of plastics. Commodity polymers, such as polyethylene, polypropylene, and polystyrene, are already produced near the well head in oil-producing countries, such as Saudi Arabia. In the industrial countries, intense research efforts are under way to develop value-added specialty polymers. Examples are the segmented copolymers presently used to mold noncorrosive plastic cars. Superior hot-melt adhesives are also formulated to bond plastic parts used in the automotive and aerospace industries. The development of plastic compact discs allows storage of enormous amounts of data. This development has already revolutionized the audio industry, and it has been predicted that by 1994 1.5 billion digital audio discs will be sold. Biopolymers are the new frontier in pharmaceuticals. They can be synthesized in a stepwise fashion using polymeric substrates (Merrifield resins) or, better, by genetic engineering. Several new biopolymers are already on the market, and many more will be developed in coming years.

Synthetic materials have become essential to the well-being of humans on this earth. The world population is ever increasing, providing the driving force for further growth of industrial polymers. The manufacture of polymers causes less environmental stress than the growing of wool, cotton, and wood, especially since we need to expand agricultural land use for food production. Plastic products are often more energy efficient than alternative materials, and past experience suggests that there are no practical limits to the extent to which synthetic polymers may serve needs in the next century and beyond.

2 Basic Polymer Principles

The unique properties of polymers are attributed to their long-chain structure. Key physical properties are directly dependent on molecular weight and molecular structure. A minimum molecular weight of 10,000 is required for the polymer to have useful physical properties. Free radical or ionic addition polymerization reactions yield significantly higher molecular weights than condensation polymerization reactions. The reason for the high molecular weights obtained in radical and ionic addition reactions ist the observed chain-growth mechanism. Each polymer chain, once intitiated, grows at an extremely rapid rate until the reaction is terminated. In contrast, condensation polymerizations proceed by a step-growth process in which each polymer chain grows at a relatively slow rate. The monomer can react only with an active end group on the growing polymer chain. The average molecular weight of vinyl polymers is on the order of over 100,000, while typical condensation polymers have molecular weights of less than 100,000. The molecular weight must be high enough to overcome the inherent brittleness of low molecular weight oligomers.

The melt viscosity of a polymeric compound, an important factor in its processability, is dependent on the weight-average molecular weight. Melt viscosity is influenced by chain length and branching, but crosslinking of polymer chains has the most pronounced influence. Linear polymers are thermoplastic, that is, they can be remelted many times, while crosslinked polymers are thermoset materials which cannot be reprocessed. Linear polymers have a degree of solubility. Crosslinked polymer networks are insoluble. A general description of the relationship of structure and properties in common industrial polymers is given in Table 2.

Table 2 Basic Polymer Structures

	Linear Chain	Network
Type of Polymerization	Polyaddition Polycondensation	Polyaddition with crosslinking Polycondensation
Supramolecular Structure	Crystalline, Amorphous	Amorphous
Processing	Thermoplastic	Thermoset
Solubility	Soluble	Insoluble

Polymers are viscolelastic materials that can show all the features of a glassy, brittle solid, an elastic rubber, or a viscous liquid, depending on the temperature and time scale of measurement. For example, a polymer may be glasslike at low temperatures and rubberlike at high temperatures. At still higher temperatures, permanent deformation occurs under load, and the polymer behaves like a viscous liquid. In an intermediate temperature range, commonly called the glass transition range or temperature (T_g), the polymer shows intermediate behavior. Above its T_g it loses some of its mechanical properties, because in its rubbery state it deforms much more readily.

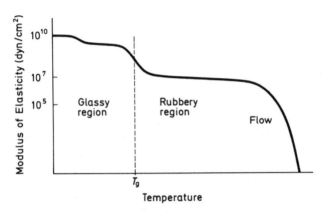

Figure 1   Modulus of elasticity versus temperature for a typical polymer. T_g = glass transition temperature

A typical modulus versus temperature curve is shown in Figure 1. In the relatively high modulus region ($\sim 10^{10}$ dyn/cm²) polymers are hard and stiff (glassy) and often brittle. Below the glass transition temperature, the midpoint of the rapidly dropping portion of the curve in Figure 1, the molecular motion is frozen and the material is capable of undergoing only small-scale elastic deformations because carbon-carbon bond angles may be stretched slightly. As the temperature is raised from below the T_g, the material becomes leathery. This state is associated with the onset of molecular motion. As the temperature increases, the material becomes more rubbery. In the relatively constant low-modulus region ($\sim 10^7$ dyn/cm²), there may be a fairly high degree of elasticity, here the molecules have considerable segmental motion. At still higher temperatures a free-flowing liquid region is observed; this is the state used for molding or extrusion. Modulus of elasticity versus temperature curves for rubber, polystyrene, and polyethylene are shown in Figure 2.

The glass transition temperatures for a group of polymers are shown in Table 3. A higher degree of chain stiffness usually gives a higher T_g, as shown in Table 4 for a series of chlorinated polyethylenes.

The melt temperature (T_m) of a polymer is especially important for processability. Most synthetic fibers are melt spun, and a T_m below 300 °C is important because significant thermal degradation of the polymer occurs above 300 °C. If the resulting fabric is to be subjected to ironing, the T_m has to be above 200 °C. In selecting a polymer for use in fibers, melt characteristics must be taken into consideration. For synthetic fibers, linear crystalline polymers with a realtively high melting point, high symmetry, and high intermolecular forces are usually used.

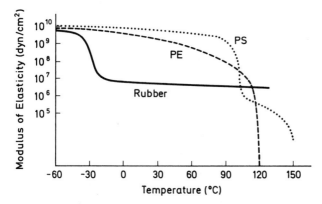

Figure 2 Modulus of elasticity versus temperature for rubber, polystyrene (PS), and polyethylene (PE)

Table 3 Glass Transisiton Temperatures (T_g) of Several Polymers

Polymer	T_g (°C)
Natural rubber	−72
Polyisobutylene	−70
Polypropylene	5
Polyvinyl acetate	29
Nylon-6	60
Polyvinyl chloride	82
Polystyrene	100
Polymethyl methacrylate	105
Polycarbonate	150
Polyvinylcarbazole	211

Table 4 Glass Transition Temperatures of Chlorinated Polyethylenes

% Chlorine	T_g (°C)
0	−50
30	−20
50	20
60	75
70	150
72	180

In contrast, elastomers should be amorphous linear polymers with high local mobility in the chain. The T_g should be ideally in the range of −50 to −80°C, and the strength of the elastomers is provided by either domain crystallinity caused by polar group interaction (polyurethane elastomers) or by introduction of a few crosslinks (olefin rubbers). A stretched elastomer should have high tensile strength and a modulus similar to those of crystalline plastics.

The wide range of end uses in plastic materials requires a variety of property combinations intermediate between those of fibers and elastomers. The most important property for mechanical applications is toughness. For electrical applications, low dielectric loss over a wide frequency range is desirable, and for engineering plastics, which are tailored to replace metals, an outstanding balance of properties is necessary. Engineering plastics must be strong, stiff, tough, abrasion-resistant, capable of withstanding wide ranges of temperature, and resistant to attack by weather and solvents.

Physical properties of polymers are also influenced by atomic interaction between chain atoms. The ability of polymers to crystallize, the flexibility of their chains, and the spacing of polar groups is of considerable importance. Ideally one differentiates between crystalline and amorphous or glassy polymers. Most industrial polymers, however, have variable degrees of crystallinity. Often the degree of crystallinity is influenced by the conditions of fabrication. Supramolecular structures also play an important role in determining the physical properties of polymers. Specific topographical aggregates can lead to crystalline regions in an otherwise amorphous polymer. This interaction can involve many chains, thereby aligning molecular segments via hydrogen bonding or other, nonbonding, interactions.

Chain flexibility in macromolecules arises from rotation around saturated chain bonds. Substitution of nonpolar groups for hydrogen in a polymer chain could lead to a loss of crystallinity. Random substitution, as in branched polyethylene, results in reduction of the size of the crystalline regions. The T_m of polyethylene decreases by 20 to 25 °C on going from linear to more branched macromolecules. Replacement of an amide hydrogen with an alkyl group has a much greater effect on the T_m, since hydrogen bonding is destroyed. N-Methyl nylons melt at considerably lower temperatures than their unsubstituted counterparts.

Regular substitution of an alkyl group into a methylene chain with retention of stereoregularity could lead to an increase in stiffness, especially with bulky alkyl groups, which increase the T_m of the polymer (see Table 5).

In olefin homopolymerization reactions involving substituted olefins, different chemical isomers can be formed depending on the type of addition (head-to-tail and head-to-head: see Figure 23). This is discussed in Chapter 7, where the more complex concept of stereoregular polymerization is also discussed in detail.

Polymers derive their usefulness from the obtainable mechanical properties, such as stress-strain (or tensile-elongation) and stress-relaxation (or tensile-time) behaviors. Amorphous polymers above their T_g show low stress values (low tensile strength) accompanied by relatively low strain (elongation). In contrast, a hard and brittle material, such as an amorphous polymer below its T_g, shows moderate tensile strength but very low elongation.

Table 5 Effect of Side-Chains on the Melt Temperature of Isotactic Polyolefins

Side Chain	T_m (°C)
—CH$_3$	165
—CH$_2$CH$_3$	125
—CH$_2$CH$_2$CH$_3$	75
—CH$_2$CH$_2$CH$_2$CH$_3$	−55
—CH$_2$—CH(CH$_3$)CH$_2$CH$_3$	196
—CH$_2$—C(CH$_3$)$_2$—CH$_2$CH$_3$	350

High molecular weight crystalline polymers are hard and tough. The crystalline regions provide high strength. Hardness is related to T_g, whereas elasticity depends upon the ability of disordered chain segments to be straightened out under the influence of stress. Elasticity is usually observed above the glass transition temperature. Highly crystalline materials are expected to show less elasticity. The impact strength of a polymer is related to its ability to absorb energy. Materials below T_g are low in impact strength. An exception is polycarbonate, which is high in impact strength at room temperature in spite of its high T_g (150 °C). This phenomenon can be explained by the energy absorption capabilities associated with rotation of the phenyl rings in the backbone of the molecules. Stiff chains (those containing large side groups or phenyl groups in the chain) tend to have a high T_g and, if they are crystallizable, tend to have high melt temperatures. Regular chains (those having stereoregular structures or simple regular structures) can crystallize most readily.

3 Raw Materials for the Polymer Industry

Today's polymers are produced predominantly from natural gas or petrochemical sources. Viable alternative raw materials for making polymers are coal and wood. Inorganic materials, which are more abundant, are not suitable, because most inorganic bonds are vulnerable to attack by both water and oxygen. Exceptions are glass, an inorganic network polymer, and organic substituted inorganic materials such as silicones and phosphazenes. The natural distribution of the elements on earth is shown in Table 6.

Table 6 Distribution of Elements on Earth

Element	%	Element	%
Oxygen	49.2	Chlorine	0.19
Silicon	25.67	Phosphorus	0.11
Aluminum	7.50	Manganese	0.08
Iron	4.71	Carbon	0.08
Calcium	3.39	Sulfur	0.06
Sodium	2.63	Barium	0.04
Potassium	2.40	Nitrogen	0.03
Magnesium	1.93	Fluorine	0.03
Hydrogen	0.87	Strontium	0.02
Titanium	0.58	Others	0.47

Almost all industrial polymers are based on carbon and hydrogen, and in addition to natural gas, oil, oilsand, and coal (depletable resources), renewable resources such as algae, wood, and other plant products should be considered for use as raw materials for polymer production. The main potential renewable resource is cellulose. Naturally occurring cellulose products are often used without chemical modification in fibers (cotton, hemp, sisal, linen), and chemically reacted cellulose products are used in fibers (rayon) and plastics (cellulose acetate). The usefulness of cellulose materials, however, is limited by their poor mechanical properties and inferior wash-and wear and soil-release properties in fiber applications. The primary source of industrial cellulose is wood, whose other component is lignin, which can be degraded to produce phenol, another useful polymer building block. The major applications of cellulose are summarized in Figure 3.

3 Raw Materials for the Polymer Industry

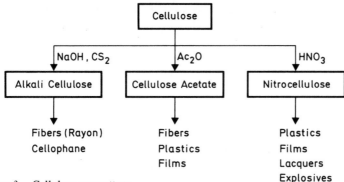

Figure 3   Cellulose use pattern

Feedstock prices are an important consideration in the plastics industries. Presently, natural gas accounts for 48 % of the total olefin production in the United States, with petroleum sources accounting for the remaining 52 %. In Europe and Japan, natural gas is not abundantly available and most olefins are produced from oil. The aromatic building blocks for industrial polymers are obtained almost exclusively from petroleum, but coal tar may play an important role in future years.

Figure 4 shows the pathways associated with the major industrial monomers and polymers derived from ethylene.

The extensive use of naphtha as the chemical feedstock promoted the generation of products from the readily available coproducts (propylene, butadiene, and aromatics).

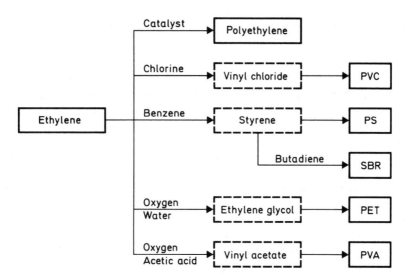

Figure 4   Monomers and polymers derived from ehtylene.
PVA = Polyvinyl acetate; SBR = styrene-butadiene rubber; PET = polyethylene terephthalate

Ethane obtained from natural gas provides up to 85 % ethylene; in contrast, cracking of naphtha produces only 30 % ethylene. The growth of the propylene-based industry during the 1960s and 1970s was due largely to the plentiful supply generated by the shift to naphtha feedstocks as well as the development of economically attractive methods for the homopolymerization of propylene. The significance of propylene as the basic building block for a number of industrial polymers is shown in Figure 5.

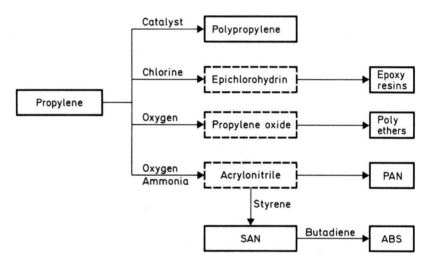

Figure 5 Monomers and polymers derived from propylene.
PAN = polyacrylonitrile; SAN = styrene-acrylonitrile copolymer; ABS = acrylonitrile-butadiene-styrene terpolymer

The aromatic BTX (benzene, toluene, xylene) fraction obtained from catalytic reformate is the third most important feedstock for industrial polymers. Benzene reacts with ethylene to provide ethylbenzene, which on dehydrogenation leads to styrene. It also reacts with propylene to give cumene, which oxidizes with air to phenol and acetone. The third important reaction of benzene is its hydrogenation to cyclohexane, which provides the intermediates for nylon-6 and nylon-6,6. Monomers and polymers derived from benzene are summarized in Figure 6.

Toluene is a basic raw material for polyurethanes (see Chapter 12) and it is also used to produce benzene by dehydroalkylation. The three isomers of xylene, the third component of the BTX fraction, are oxidized to give the corresponding dicarboxylic acids, which are used as intermediates for the manufacture of condensation polymers. Oxidation of *o*-xylene produces phthalic anhydride, which is widely used in the manufacture of unsaturated polyesters; oxidation of *m*-xylene yields isophthalic acid, a monomer for high-temperature polymers; and oxidation of *p*-xylene affords terephthalic acid, the basic raw material for polyethylene and polybutylene terephthalate.

3 Raw Materials for the Polymer Industry 25

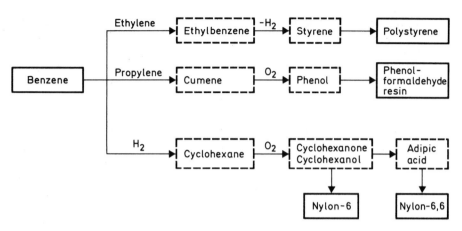

Figure 6 Monomers and polymers derived from benzene

Of the 50 highest volume chemicals used in the chemical industry, four are based on methane, the main component of natural gas. These four are urea, methanol, formaldehyde, and acetic acid. Reaction of methane with ammonia and oxygen leads to hydrogen cyanide, which is required for the manufacture of acrylates, methacrylates, and hexamethylene diamine. Another important use of methane is its conversion into synthesis gas, a mixture of hydrogen and carbon monoxide obtainable from natural gas or petroleum. Synthesis gas, called water gas when it is derived from coal, is the basis for a number of important industrial processes (see Figure 7).

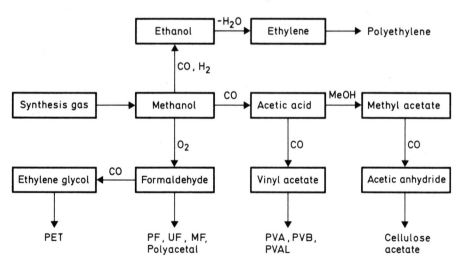

Figure 7 The derivation of monomers and polymers from synthesis gas.
PET = Polyethylene terephthalate; PF, UF, MF = phenol-, urea- and melamine-formaldehyde resins; PVA = polyvinyl acetate; PVAL = polyvinyl alcohol; PVB = polyvinyl butyral

The rapid increase of the price of oil has again focused attention on the plentifully available raw material coal. The direct cracking of coal to obtain ethylene is not yet feasible. However, water gas, used widely in polymerization processes (see Figure 7), is readily available from coal. The carbon monoxide and hydrogen from coal can be combined to give methane, one of the approaches to generate a substitute for natural gas. Water gas may also be converted by the Fisher-Tropsch reaction into a petroleumlike material.

Another basic raw material for polymer intermediates which can be readily obtained from coal is acetylene. Acetylene is obtained by converting coke into calcium carbide and treating the latter with water. Attempts to develop lower energy routes to acetylene have failed so far, and acetylene remains a relatively unimportant raw material for the polymer industry. However, pathways to possible applications are diagrammed in Figure 8.

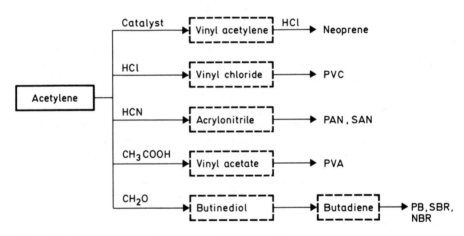

Figure 8 Monomers and polymers derived from acetylene.

4 Production of Industrial Polymers

Industrial polymers are produced in bulk, in solution, or in suspension. The bulk polymerization process is commonly used for the manufacture of linear condensation polymers, because step-growth polymerization is only mildly exothermic and most of the reaction occurs when the viscosity of the mixture is still low enough to allow completion of reaction. Some addition polymers are also produced by the bulk polymerization method. A schematic diagram of the continuous bulk polymerization of polystyrene is shown in Figure 9.

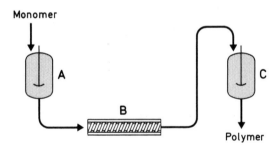

Figure 9 Continuous bulk polymerization of polystyrene.
A prepolymerization reaction (30 to 50 % conversion); B polymerization screw (90 to 95 % conversion); C vaporizer for removal of residual monomer

The highly exothermic free radical initiated polymerization of vinyl monomers is usually conducted in solution or suspension. Often the monomer is used as solvent, because it is easy to separate the insoluble polymer from excess monomer. Inert hydrocarbon solvents are used in the stereospecific homopolymerization of olefins using Ziegler-Natta catalysts. In suspension polymerization, a liquid monomer is suspended as liquid droplets and the resultant polymer is obtained as a dispersed solid phase. Initiators that are soluble in the liquid monomer phase are used, and the polymer is obtained in the form of pearls or beads. A batch suspension polymerization process for polyvinyl chloride is shown diagrammatically in Figure 10.

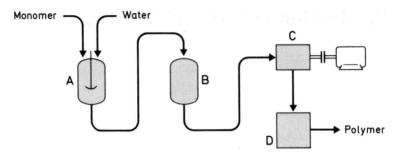

Figure 10 Batch suspension polymerization of polyvinyl chloride
A reactor; B slurry storage; C polymer centrifuge; D dryer

Emulsion polymerization is very similar to the suspension polymerization method. The reaction occurs in an emulsion formed by adding a surfactant or soap to the monomer and is initiated in the aqueous phase. The particle size is smaller, and a narrower molecular weight distribution can be achieved. A typical batch emulsion polymerization process is outlined in Figure 11.

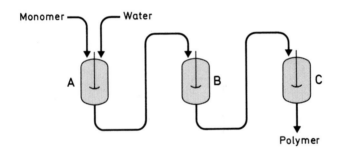

Figure 11 Batch emulsion polymerization of acrylic polymers
A preemulsification tank for preparation of monomer emulsion; B reactor; C holding tank for polymer emulsion

A disadvantage of both suspension and emulsion polymerization, which are aqueous two-phase processes, is that the polymer must be washed and dried and may become contaminated with emulsifier.

In condensation polymerization processes employing dicarboxylic acid chlorides as comonomers, the interfacial polymerization method is used. The reaction occurs at the interface of water and an immiscible organic solvent. In the production of polyamides the dicarboxylic acid chloride is added in an inert organic solvent, and the diamine and a strong base, serving as a hydrogen chloride scavenger, are dissolved in the aqueous phase.

Some special polymers are produced by conducting further chemical reactions on macromolecules. For example, polyvinyl acetate is hydrolyzed to polyvinyl alcohol, and the latter is further reacted with aldehydes to give polyacetals (see Chapter 11.) Also, nitration and sulfonation of styrene polymers are widely used to produce ion-exchange resins (see Chapter 19).

Whereas large volume thermoplastics are produced in multimillion-pound plants, most thermosetting plastics are manufactured from liquid monomers in medium-sized production facilities. Polyurethane foams, for example (see also Chapter 12), are typically poured by mulitcomponent machines at rates of over 100 lb/min in a continuos process. The reagents, including an isocyanate, a macroglycole, a blowing agent, and a surfactant, are metered and mixed together continuously and are discharged while the mixture polymerizes and expands to a cellular mass. the foam moves on a long conveyor and is cut into buns that range in width from about 56 to about 86 in and are typically 30 to 40 in high. Rigid polyurethane foam laminates or composites, used in the insulation of industrial buildings, are continuously manufactured in a similar manner. A diagram of this process is shown in Figure 12. The chemical ingredients are preblended to form three component parts, which are stored in tanks, from which they are pumped in precisely metered quantities and at controlled temperatures to the foam mixing head. The dispensed chemicals form a bank behind the metering rolls, and the foam rises and hardens in the restraining conveyor section, which is usually enclosed in an oven. After it emerges from the oven, the product is trimmed and cut to length.

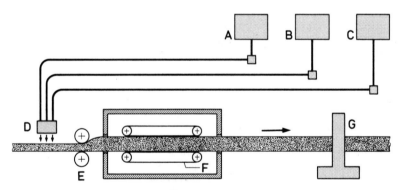

Figure 12 Process for manufacture of polyurethane foam laminates
A, B, C tanks for isocyanate, polyol, catalyst, blowing agent and surfactant; *D* mixing head; *E* metering roll; *F* restraining conveyor; *G* saw

Most plastic materials can be mixed with fibers or fillers to provide an almost unlimited range of formulations to meet most engineering applications. Examples of fibers for reinforcement of plastic materials (see also Chaper 15) include glass, carbon, boron, and aramid fibers, and naturally occurring minerals are often used as fillers. The continuous mixing or compounding of plastic materials is usually conducted in Banbury-type mixers or extruders, but discontinuous processes using a ribbon blender are also sometimes used.

5 Processing of Industrial Polymers

The processing of polymers is determined largely by the rheological properties of the macromolecules. Thermoplastic materials flow at elevated temperatures, which allows the use of a wide variety of molding processes. In contrast, thermoset polymers undergo further crosslinking reactions under processing conditions and rather rapidly lose the ability to flow. Thermosetting materials are processed mainly by compression molding.

Thermoplastic polymers are supplied in the form of pellets, granules, flakes, and powder. Most molding processes involve the melting of the polymer, followed by the application of pressure to force the molten material into a mold cavity or through a die. The molten plastic is subsequently cooled to allow it to harden. Some materials require postcuring at elevated temperature. Heat and pressure are also used with thermosets only in this case the heat serves to cure the polymer. In liquid casting, liquid monomers are used, and the polymer is produced in the mold. A typical example is the reaction injection molding (RIM) process, which is used extensively in the automotive industry. The major methods of processing industrial polymers are summarized in Table 7.

Table 7 Methods of Processing Polymers

Thermoplastics	Thermosets
Injection molding	Compression molding
Extrusion	High-pressure lamination
Blow molding	Reaction injection molding
Calendering	Reinforced plastic molding

The key methods employed for molding thermoplastic materials are injection molding and extrusion. The injection molding process was introduced by Hyatt in 1872 to mold celluloid (cellulose nitrate). In this process, the solid polymeric particles are fed through a hopper into a heated extruder barrel, where the molten material is transported into a relatively cool mold by rotating the screw followed by the plunger action of the nonrotating screw. A diagram of an injection molding machine is shown in Figure 13.

Extrusion is employed to form thermoplastic materials into continuous sheeting, film, tubes, rods, and cable. The method is similar to the one used in injection molding

5 Processing of Industrial Polymers

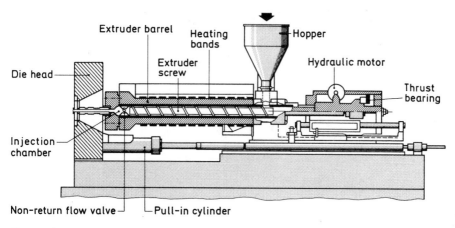

Figure 13 Injection molding machine

except that the molten polymer is forced through a small opening or die having the shape desired in the finished product. (The diagram of an extruder is shown in Figure 14.) The extrudate is fed onto a coveyor belt and through a water through, where it is cooled.

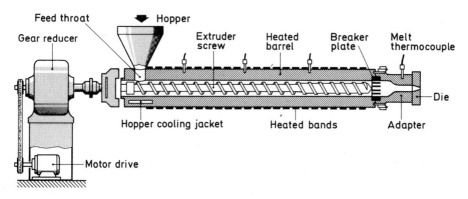

Figure 14 Single-screw extruder

In the case of wire and cable coating, the thermoplastic is extruded around a continuing length of wire or cable, which also passes through the extruder die. The coated wire is wound on drums after cooling. A similar process, called blow molding, is widely used to produce bottles and other hollow plastic products. Blow molding involves the melting of the thermoplastic polymer into a tubelike shape, much like trapping a balloon inside a mold, inflating it so that it contacts to the inside walls of the mold cavity, and ejecting the solidified finished piece when it is sufficiently cooled. A blow molding setup in which a rotating horizontal table is used in shown in Figure 15.

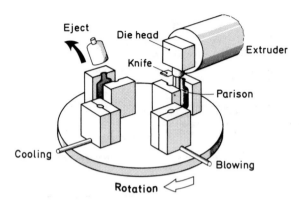

Figure 15 Blow molding

Large tubular film extruders are used in the extrusion or blowing of plastic film. In this process a thin-walled tube is extruded upward and is cooled with air. Air is also introduced through the center of the die to inflate the tube and cause it to expand. The film is flattened as it passes through converging planes and between a pair of pull rolls (Figure 16).

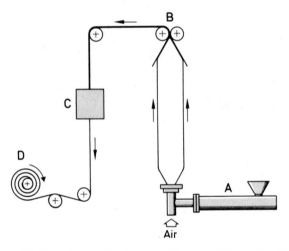

Figure 16 Extrusion of plastic film. *A* Extruder; *B* Nip rolls; *C* Slitter; *D* Winder

Calendering is another method used to process thermoplastics into film and sheeting and to apply a plastic coating to textiles. In calendering, the molten thermoplastic is passed between a series of three or four large, heated revolving rollers, which squeeze the material between them into a sheet or film. A similar technique is used for the application of a plastic coating to metal, wood, paper, fabric, etc. The diagram of a typical coating machine is shown in Figure 17.

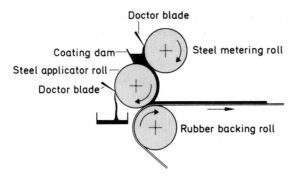

Figure 17 Coating machine

Compression molding is the most common method of forming thermosetting materials. The material is placed in a mold and simply squeezed into the desired shape by application of heat and pressure. Preheated plastic molding powder or preforms, often mixed with fillers, is added to the preheated open mold cavity, as shown in Figure 18a. The mold is then closed, pressing down on the plastic and causing it to flow throughout the mold (Figure 18b). High-pressure laminating is a similar process using high heat and pressure.

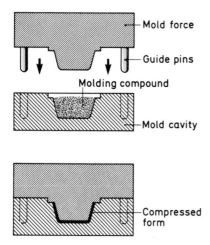

Figure 18 Compression molding. (a) mold open, (b) mold closed

A variation of this method is reaction injection molding (RIM), which so far has been used primarily for molding polyurethane elastomers or foams. In this case, high pressure pumps are used to bring two reactive streams together under high pressure in a mixing chamber. The resulting mixture is injected under low pressure into the mold where the reaction occurs. The use of liquid polymers or monomers in so-called casting processes has the advantage that pressure is not required.

Most high-volume reinforced plastic processes in use today involve the application of heat and pressure. For example, reinforced thermoset plastics are placed between matching heated molds and held under pressure until the polymer hardens into a solid (sheet molding compound). In another method, the polymer, fillers, and catalyst are mixed together into a puttylike mass (bulk molding compound).

Cellular polyurethanes can be produced in continuous or semicontinuous fashion from liquid monomers. Most flexible foam and some rigid foam is produced on continuous bun or block lines. Fabrication of semicontinuous or continuous laminate products is one of the more important methods used in the production of rigid foam insulation boards. A third method of application of rigid polyurethane foam involves spraying liquid monomers onto a suitable surface. This method is widely used for roof and tank insulation.

Synthetic fibers are produced mainly by melt spinning techniques, but dry or solvent spinning methods are also used. In the latter case, polymer solutions are utilized. These techniques are discussed in Part III, Chapters 14 and 15.

6 Recycling of Industrial Polymers

The ecological disposal of plastic waste products is of considerable concern. Discarded plastics are highly visible as litter in the marine environments and as an accumulating portion of solid waste in landfills. Proposed methods of coping with this problem include reduction of waste, formulation of degradable plastics, conversion of waste directly into useful products, thermal recycling, recycling of waste products into raw materials (chemical recycling), and reuse of plastic waste. While all the approaches are different, they have common problems – collection and separation, cost of recycling, and identification of markets for the recovered products.

Polymers account for 7 % (by weight) of the municipal solid waste generated in the United States each year from household, commercial, and industrial sources. Since most plastic materials are mutually incompatible, workable technologies have to be developed to separate and sort mixtures of plastic waste products. Automated sorting systems are under development. Contamination is a big problem. For example, PET bottles cannot tolerate PVC. Mixtures of LDPE, PVC, and PS become difficult to process at levels above 50 % PE. Chlorination of PE waste produces a product that may be useful as a low grade elastomer. When plastics cannot be sorted by type – as, for example, multilayer bottles or coextruded film – the mixture of such products can be used in non-load-bearing structural applications, like park benches.

Many of the large producers of plastics have constructed recycling facilities, or they have formed joint ventures to undertake the recycling. For example, Occidental Chemical is building the largest recycling plant for postconsumer plastics in Dallas, Texas. This plant will have the capacity to recycle 40 million pounds of plastic containers per year. The plant will accept HDPE, PET, and PVC bottles and other plastics. Occidental Chemical plans to sell "EcoVinyl," a PVC brand containing 25 % recycled PVC.

The recycling of thermoplastic waste is relatively simple, provided economic collection processes exist. In 1991, already 20 % of U. S. households had curbside collection of recyclables. The introduction of bottle bills has created an easy collection system for used containers. In-plant generated waste can be ground and mixed with the virgin polymers. In Europe in 1990 about 900,000 metric tons of thermoplastic waste materials were recycled.

Processes for PET and HDPE recycling are well under way, because bottles of these types are easily identified visually and picked from waste streams. In the United States in 1990 about 200 million pounds of PET was recycled. This amount corresponds to approximately 25 % of the PET used in soft drink bottles. For example, Coca Cola and Pepsi Cola are already using recycled PET in their soda bottles. Recycled PET is also used in the manufacture of polyester polyols, in polyurethane rigid foam, and in the manufacture of polyester staple fiber. Also about 100 million pounds of HDPE was recycled in 1990. Shell, Exxon, and Texaco are including postconsumer HDPE in their motor oil bottles. Exxon developed a HDPE grade that can accept higher levels of recycled plastic without sacrificing bottle strength.

The recycling of other plastics is also in progress. About 30 to 50 million pounds of polypropylene was recycled in the United States in 1990. Almost all the polypropylene came from automotive battery casings. The use of plastic materials in the automotive industry was about 300 pounds per car in 1990. Fiber reinforced polymer based materials are expected to be used increasingly for exterior panels. Recycling of plastic materials from discarded automobiles again would be relatively easy. In the automotive industry in Europe already polypropylene containing recycled polymer is used in the molding of parts. Major automotive companies have developed processes for the chemical hydrolysis of flexible polyurethane foam cushions with superheated steam to produce diamines and polyols. An even simpler method involves the conversion of all polyurethane products – flexible foam seating as well as elastomer products used in energy-absorbing bumpers, fascia parts, fenders, and exterior panels – into useful polyols by heating with glycol mixtures. Polyvinyl chloride fabrics and skins can also be recycled by chemical treatment processes. The recycled PVC products can be reused in plastic pipe applications.

The recycling of PS has been affected by the decision of McDonald's to stop using polystyrene clamshell packaging in its restaurants. The majority of the 20 to 25 million pounds of recycled polystyrene produced in the United States in 1990 came from food packaging. Eight U. S. polystyrene producers formed the National Polystyrene Recycling company. Actually, polystyrene's recyclability makes it the material of choice for food trays and cups now made from wax-coated paper. Wax-coated paper cannot be recycled, which adds to the landfill headache. Apparently the decision by McDonald's to stop using PS was politically motivated rather than rational.

The use of mixed plastic waste as substitutes for sand in the formulation of concrete for bridges and as a binder for sand and gravel in bricks is under development. This approach does not require the separation of the plastics. Waste processing plants to achieve separation of metals, glass, plastics, and paper are relatively simple. A flowchart of a typical waste processing plant is outlined in Figure 19.

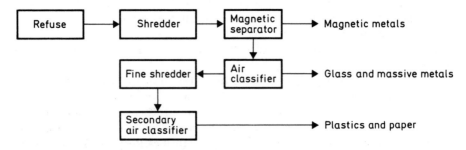

Figure 19 Outline of a waste-processing plant

Pyrolysis or thermal depolymerization processes are also feasible. Polymethyl methacrylate can be readily depolymerized to yield 100 % monomer. Polytetrafluoroethylene also gives high yields of monomer upon pyrolysis. In contrast, only 40 % styrene monomer can be obtained in the thermal degradation of polystyrene. Thermal degradation processes are not useful for polyvinyl acetate and polyvinyl chloride because removal of side groups rather than chain scission is observed. It was proposed by some researchers that the mixed plastic products be added to a feedstock stream in an oil refinery. The separation processes in the refinery recover the depolymerized chemicals.

Advanced incineration techniques can be used to exploit the high energy value of plastics (comparable to coal) to generate steam for the production of electricity. Both the pyrolysis and the incineration methods provide a continuous renewal of resources and energy. In pyrolysis, materials go from feedstock to plastic to product and back to feedstock. With incineration, one goes from feedstock to plastic to product and, through incineration, energy for heat and electricity is recovered. Incineration seems to be at present the most economical method of removing plastic waste from the environment. Recycling requires energy for the collection, sorting, and cleaning of the plastics, and it is economically viable only if the cost of the recycled materials is lower than the cost of the virgin materials. However, public preference for recycled plastics may force the plastic producers to become recyclers in order to compete effectively.

Another solution to the plastic litter problem would be the formulation of degradable plastics. The large plastic producers Dow Chemical and Union Carbide produce copolymers of polyethylene with about 1 % of carbon monoxide for use in light degradable six-pack beverage rings. Biodegradable polymers based on cornstarch, cellulose, lactic acid, and proteins are under development. For example, poly(3-hydroxybutyrate-3-hydroxyvalerate) (PHBV) is produced by ICI. PHBV behaves like polypropylene but it is readily degradable by microorganisms (see Chapter 14).

A new technology to convert scrap rubber into useful monomers by treatment with ethylene in the presence of a metathesis catalyst has recently been developed.

$$CH_2=CH_2 \ + \ \left[\begin{array}{c} | \\ \diagup \diagdown \diagup \diagdown \\ \end{array} \right]_n \ \longrightarrow \ \begin{array}{c} | \\ \diagup \diagdown \diagup \end{array} \ + \ \text{oligomers}$$

This process may provide a solution to the accumulation of used automotive tires.

Part II
Addition Polymers

7 Polyolefins

Introduction

Polyolefins are the volume leaders in industrial polymers. They are based on low-cost petrochemicals, or natural gas and the required monomers are produced by cracking or refining crude oil. The most important monomers are ethylene, propylene, and butadiene. For example, 50 % of the 37.5 billion lb of ethylene produced in the United States in 1990 was used in the production of polyethylene; 15 % in the production of vinyl chloride, the monomer for polyvinyl chloride; 20 % in the production of ethylene oxide, another polymer building block; and 10 % in the production of styrene monomer. Benzene, the other major building block for styrene monomer, is extracted from the catalytic reformate made in oil refineries but is also obtainable from pyrolysis gasoline mad in steam-cracking olefin plants and from the dealkylation of toluene.

The outstanding growth rate of the four major polyolefins, polyethylene, polyprophylene, polyvinyl chloride (PVC), and polystyrene is based strictly on economics. They are by far the least expensive industrial polymers on the market. In Table 8 the volume – price relationship of several polyolefins is shown. Since polypropylene is a relatively new product, it has not yet reached the projected volume based on price. Specialty polymers, such as the fiber-forming polyesters (10,184,000 tons in 1990; $ 0.80/lb) and nylon (4,715,000 tons in 1990; $ 1.35/lb), command a higher price based on their performance. Other high-performance products are engineering thermoplastics such as acetal, polycarbonate, polyphenylene oxide, polyphenylene sulfide, and polysulfone, which are priced in the order of $ 1.00 to 5.00/lb but whose use, of course, declines as the price increases.

Table 8 Volume-Price Relationship of Major Polyolefins

Product	1990 World Consumption* (thousand tons)	U.S. Price ($/lb)
Polyethylene	30,454	.31–.33
Polypropylene	6,131	.35–.47
PVC	18,180	.26–.28
Acrylics	2,500	.75

* Three major world markets: United States, Western Europe, and Japan.

The manufacture of the leading thermoplastics (LDPE, HDPE, PP, PS, an PVC) is a world wide endeavor, with major producers located in all industrialized countries. The companies listed in Table 9 are all multinationals, with plants and subsidiaries in major market areas. The table does not include polymers for synthetic fiber production.

Table 9   Output by Major World Plastics Producers 1980 (in thousands of tons)

Company	Major Thermoplastics*	Other Plastics	Urethane Chemicals	Total
Dow	2 546	298	250	3 094
BASF	2 507	300	264	3 071
Shell	2 821	23	62	2 906
Hoechst	2 469	49	8	2 526
ICI	2 221	49	150	2 420
Union Carbide	2 014	–	329	2 343
Montedison	1 795	270	138	2 203
Monsanto	1 522	477	–	1 999
Solvay	1 615	–	–	1 615
Hüls	842	331	–	1 173
DuPont	900	87	155	1 142
Mitsubishi	1 010	121	10	1 141
Rhône-Poulenc	940	60	134	1 134

* LDPE, HDPE, PP, PS PVC.

Polymerization of olefins is a general reaction, and in addition to the four major products a wide variety of other polyolefins are commercially available (see Table 10). In all polyolefins the repeating unit in the macromolecule is identical with the monomer.

Table 10   Polymerization of Olefins

$$RCH=CH_2 \longrightarrow -[CH(R)-CH_2]_n-$$

R	Name	R	Name
H	Polyethylene (PE)	CN	Polyacrylonitrile (PAN)
CH$_3$	Polypropylene (PP)	COOH	Polyacrylic acid
C$_6$H$_5$	Polystyrene (PS)	COOR	Polyacrylates
pyridyl	Polyvinyl pyridine	Cl	Polyvinyl chloride (PVC)

7 Polyolefins

Four types of polymerization processes for polyolefins are known: free radical initiated chain polymerization, anionic polymerization, cationic polymerization, and organometallic initiation (Ziegler-Natta catalysts). By far the most extensively used process is the free radical initated chain polymerization. The relative importance of the various polymerization processes in industrial polymer production is shown in Figure 20.

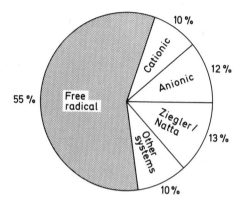

Figure 20 Relative importance of polymerization processes

Most large-volume polymers have historically been produced by free radical initiated polymerization. The more recent development of stereoregular polymers using Ziegler-Natta catalysts has added a new dimension of sophistication, and it is anticipated that these catalysts will play a more important role in coming years. The emergence of linear low-density polyethylene is a good example. Anionic polymerization is used mainly in the copolymerization of olefins (see the section on styrene-butadiene elastomers in Chapter 8), and cationic polymerization is exclusively used in the production of butyl rubber (see Chapter 8). The applicability of various types of catalysts is shown in Table 11.

Table 11 Applicability of Various Types of Initiators in the Polymerization of Olefin Monomers

Monomer	Free Radical	Anionic	Cationic	Ziegler-Natta
Ethylene	+	–	+	+
Propylene	–	–	–	+
Butadiene	+	+	–	+
Styrene	+	+	+	+
Vinyl chloride	+	–	–	+
Tetrafluoroethylene	+	–	–	+
Vinyl ethers	–	–	+	+
Vinyl esters	+	–	–	–
Acrylic esters	+	+	–	+
Acrylonitrile	+	+	–	+

The mechanism of the free radical chain polymerization is summarized for polyethylene in Figure 21. The most commonly used initiators include dibenzoylperoxide, di-t-butylperoxide, and azodiisobutyronitrile. The three basic steps involved in the free radical chain polymerization are initiation, propagation, and termination. The slow rate-determining step is usually the initiation of polymerization, and the rate of radical transfer increases with temperature.

Initiation

R• + C_2H_4 $\longrightarrow$ RC_2H_4•

Propagation

RC_2H_4• + C_2H_4 $\longrightarrow$ $RC_2H_4C_2H_4$•

Radical Transfer

Intramolecular:

$RC_2H_4C_2H_4$• $\longrightarrow$ $R\overset{\bullet}{C}H(CH_2)_2CH_3$

Intermolecular:

RCH_2• + $R'(CH_2)_n R''$ $\longrightarrow$ RCH_3 + $R'\overset{\bullet}{C}H(CH_2)_{n-1} R''$

Termination

By combination:

RCH_2• + $R'CH_2$• $\longrightarrow$ RCH_2CH_2R'

By disproportionation:

RCH_2CH_2• + $R'CH_2CH_2$• $\longrightarrow$ $RCH=CH_2$ + RCH_2CH_3

Figure 21 Free radical initiated chain polymerization of polyethylene

Most polyolefins are manufactured by free radical initiated chain polymerization processes, and a number of different processes are used in industry. Basically homogeneous solution polymerization and heterogeneous suspension polymerization processes are carried out. The homogeneous bulk polymerization process is economically the most attractive. However, there are problems associated with the heat of polymerization, increases in viscosity, and removal of the monomer. Nevertheless, this method is being used for the manufacture of polyvinyl chloride, polystyrene, and polymethyl methacrylate. Solution polymerization can be used for the manufacture of polyethylene, polypropylene, and polystyrene, but far the most widely used process for polyvinyl chloride and polystyrene is suspension polymerization. In this process the monomer is suspended in water, and sometimes an emulsifier is added (emulsion polymerization).

7 Polyolefins

In addition to the standard chain polymerization processes, ring-opening polymerization has been used in the manufacture of special bicyclic olefins. A commercial example is the production of polynorbornene.

The norbornene monomer is obtained readily by the Diels-Alder reaction of ethylene with cyclopentadiene, and the polymer (1,3-cyclopentylene vinylene) is an amorphous product with a softening point of 35 °C. The expanded structure of polynorbornene powder facilitates rapid incorporation of aromatic liquids or oils. For example, aromatic petroleum fractions are compatible to ten times their weight. Since polynorbornene and most oils are lighter than water, the floating polymer and oil mixture can be recovered easily from aqueous oil spills.

The polyolefins derive their physical properties from the arrangement or entanglement of the atoms in the chain molecules. Branching caused by radical transfer influences the physical properties as well as molecular distribution. The thermodynamically favored structure for polyethylene is the extended or *trans* conformation, but the bent or skew conformation contributes also to the overall chain configuration. (See Figure 22). Polyethylene is a crystalline polymer with a melting range of 130 to 138 °C. In contrast, polytetrafluoroethylene (teflon), in which all hydrogens are replaced by fluorine, is a much stiffer molecule, because the bent or skew conformation is sterically prohibited. Teflon melts at about 325 °C.

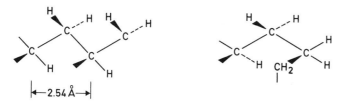

Figure 22 Polyethylene conformations. Extended (*trans*) conformation. Bent (skew) conformation

In the case of substituted olefins, polymerization can occur either by head-to-tail or by head-to-head polymerization (Figure 23). In general, head-to-tail polymerization is overwhelmingly favored, and head-to-head polymers have to be synthesized by special

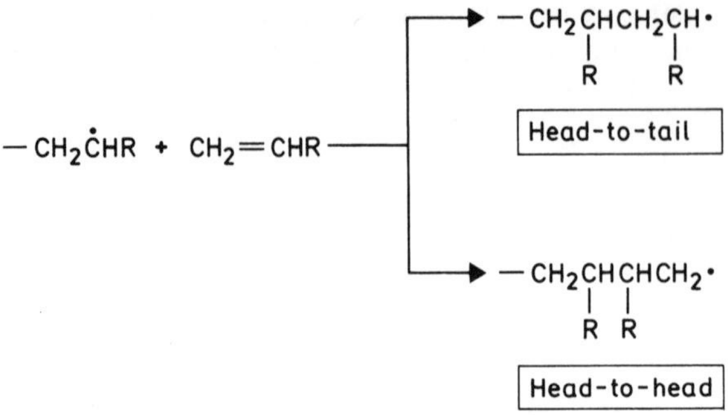

Figure 23 Repeat units in substituted polyolefins

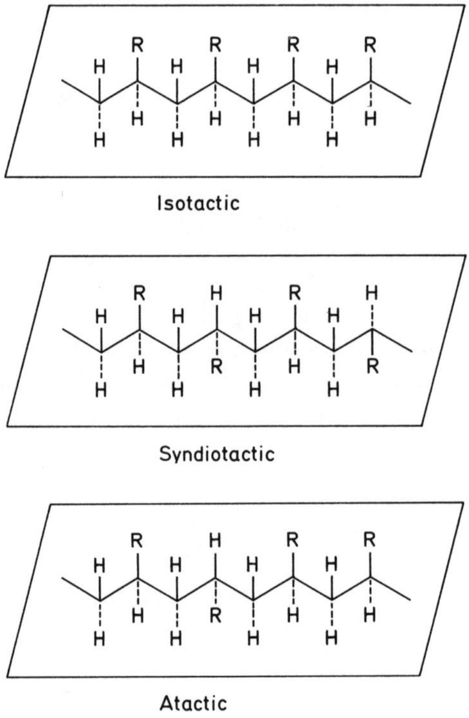

Figure 24 Structures for substituted polyolefins

methods. Ziegler and Natta received the Nobel Prize for the discovery that certain organometallic coordination compounds catalyze the formation of stereoregular polymers. The obtainable structures for substituted polyolefins are shown in Figure 24.

7 Polyolefins

The order obtainable in the macromolecule is called *tacticity*. An atactic polymer is a random polymer that is amorphous or rubbery. In the crystalline isotactic polypropylene (m. p. 170 °C) all methyl groups are above or below the horizontal plane, and in the stereoregular syndiotactic polypropylene the methyl groups alternate above and below the horizontal plane. The strain from the van der Waals overlap of the eclipsed 1,3-methyl groups is relieved by the fact that the linear chains assume a regular helical conformation. Isotactic polypropylene is never perfectly stereoregular even in the more highly crystalline grades. The degree of isotacticity may vary from 88 to 97 %.

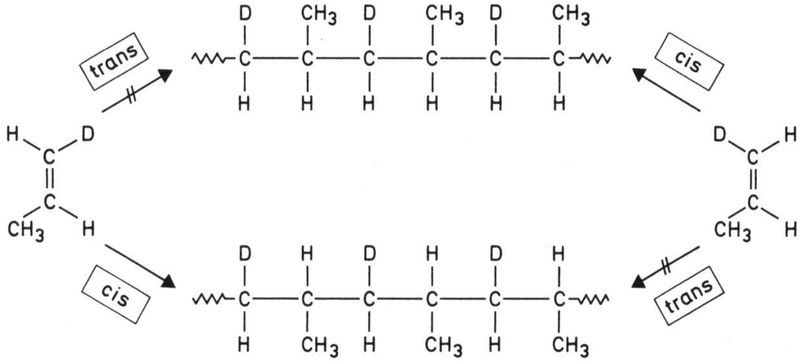

Figure 25 Stereochemistry of polypropylene chain growth

The structures of the catalytic center in the Ziegler-Natta catalysts are still unknown. Ziegler used $TiCl_4$ and aluminum alkyls in his pioneering work, and Natta used crystalline $TiCl_3$, obtained by the reduction of $TiCl_4$ with hydrogen, suspended in hydrocarbons in combination with trialkylaluminum or dialkylaluminum chloride. In this manner the isotacticity of polypropylene could be improved from 30 % to about 90 %. The growth of the polymer chain is caused by insertion and *cis* addition of the double bond. The stereochemistry of the addition to the double bond was elucidated using deuterated propylene (see Figure 25). If the olefin has two different substituents on the carbon atoms, as for example in pentene-2, the polymer can exhibit two diisotactic and two disyndiotactic forms (see Figure 26). The nomenclature is consistent with that used for organic neighboring diastereoisomers with two optically active carbon atoms.

In diolefins, such as butadiene, 1,2- as well as 1,4-polymerization can occur (Table 12). All stereoregular polymers have been synthesized using Ziegler-Natta catalysts. Stereospecificity in all cases was in the order of 99 %. 1,4-*cis*-Isoprene is the synthetic equivalent of natural rubber.

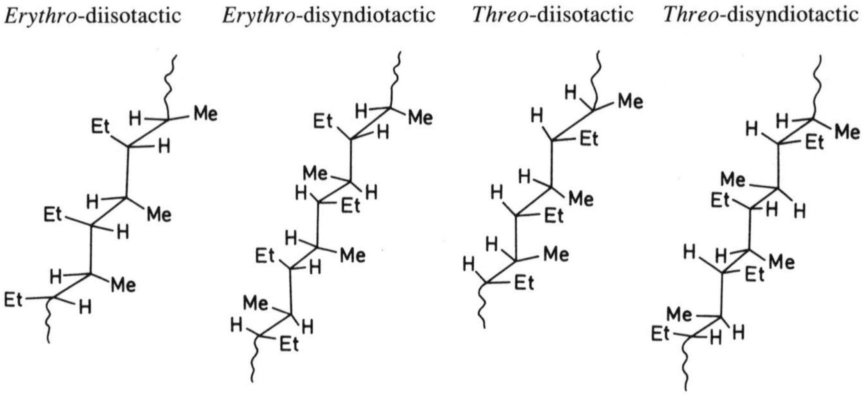

Erythro-diisotactic   *Erythro*-disyndiotactic   *Threo*-diisotactic   *Threo*-disyndiotactic

Figure 26 Polypentene-2 conformations

Table 12 Stereoregular Polybutadienes

Type of Polymer	Structure	M.p. (°C)	Catalyst
1,2-Isotactic	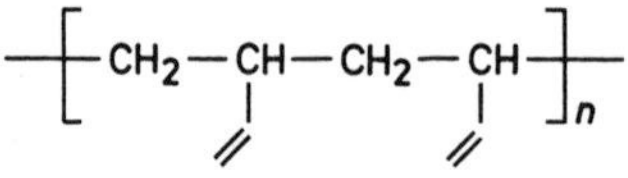 $-[CH_2-CH-CH_2-CH]_n-$ with vinyl groups	126	Cr(NCPh)$_6$ AlEt$_3$ (Al/Cr > 10)
1,2-Syndiotactic	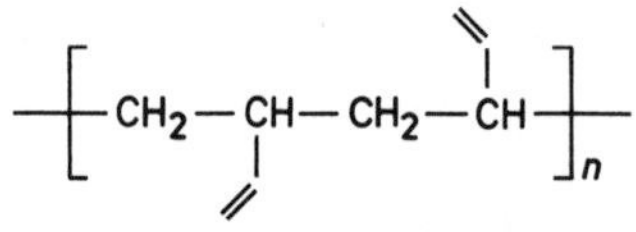 $-[CH_2-CH-CH_2-CH]_n-$ with vinyl groups	156	Cr(NCPh)$_6$ AlEt$_3$ (Al/Cr = 2)
1,4-*cis*	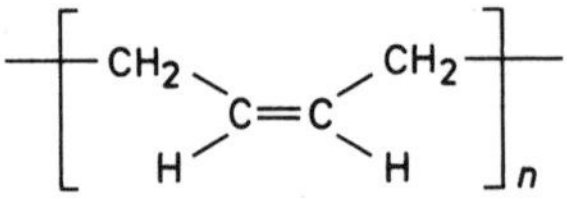 $-[CH_2-C(H)=C(H)-CH_2]_n-$	1	Cobalt chelate AlEt$_3$ Chlorobenzene
1,4-*trans*	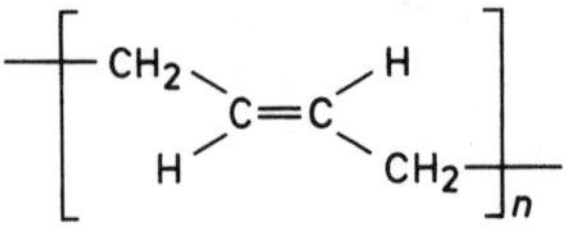 $-[CH_2-C(H)=C(H)-CH_2]_n-$	146	VCl$_3$, AlEt$_3$

Anionic chain polymerization is favored if electron-withdrawing groups are attached to the double bond (styrene, acrylonitrile, vinylidene dicyanide, etc.). Catalysts include sodium naphthanate, sodium amide, lithium alkyls, and others. The mechanism of the anionic polymerization of styrene using potassium amide in liquid ammonia is depicted in Figure 27.

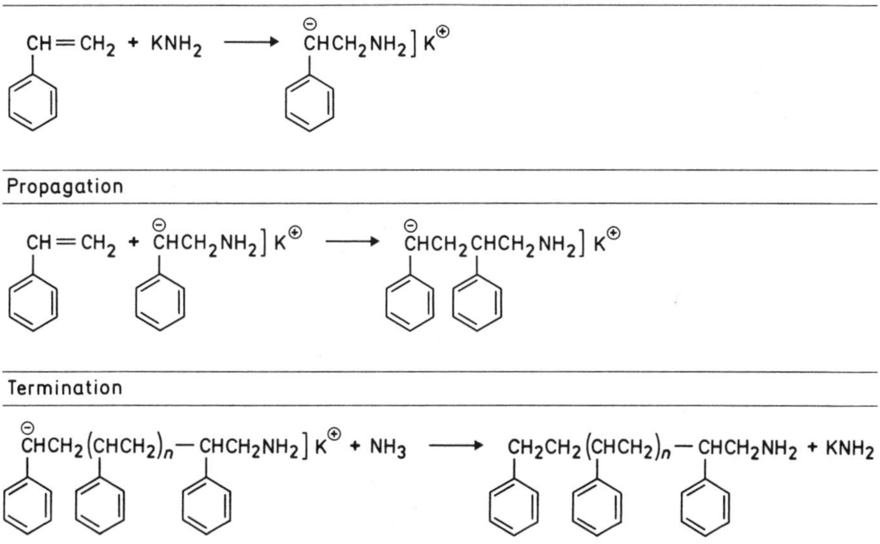

7 Polyolefins

$$CH=CH_2 + KNH_2 \longrightarrow [\overset{\ominus}{C}HCH_2NH_2] K^{\oplus}$$
(with phenyl groups on the CH carbons)

Propagation

$$CH=CH_2 + [\overset{\ominus}{C}HCH_2NH_2] K^{\oplus} \longrightarrow [\overset{\ominus}{C}HCH_2CHCH_2NH_2] K^{\oplus}$$

Termination

$$[\overset{\ominus}{C}HCH_2(CHCH_2)_n-CHCH_2NH_2] K^{\oplus} + NH_3 \longrightarrow CH_2CH_2(CHCH_2)_n-CHCH_2NH_2 + KNH_2$$

Figure 27 Mechanism of anionic chain polymerization of polystyrene

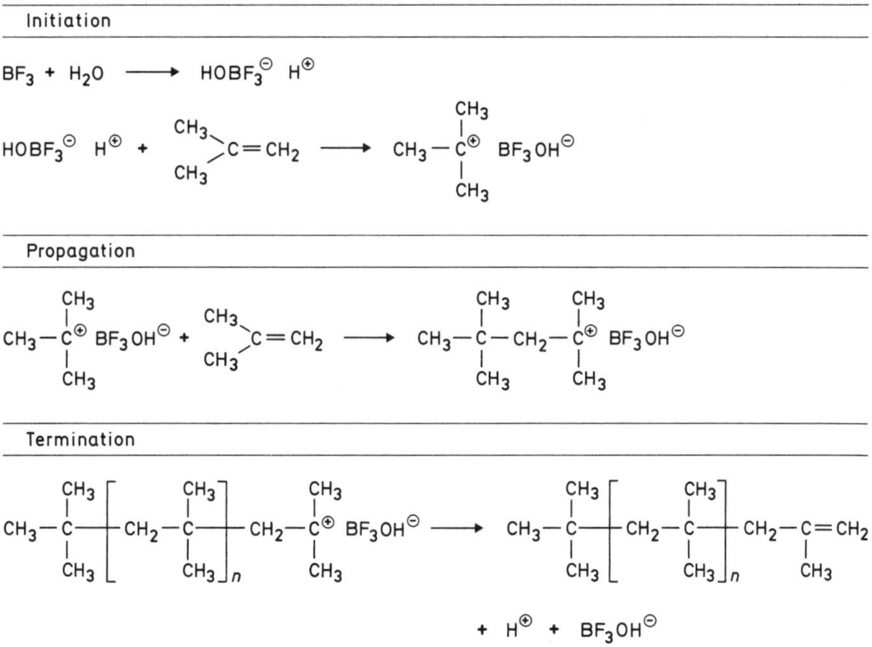

Initiation

$$BF_3 + H_2O \longrightarrow HOBF_3^{\ominus} H^{\oplus}$$

$$HOBF_3^{\ominus} H^{\oplus} + \underset{CH_3}{\overset{CH_3}{>}}C=CH_2 \longrightarrow CH_3-\underset{CH_3}{\overset{CH_3}{\underset{|}{C^{\oplus}}}} BF_3OH^{\ominus}$$

Propagation

$$CH_3-\underset{CH_3}{\overset{CH_3}{\underset{|}{C^{\oplus}}}} BF_3OH^{\ominus} + \underset{CH_3}{\overset{CH_3}{>}}C=CH_2 \longrightarrow CH_3-\underset{CH_3}{\overset{CH_3}{\underset{|}{C}}}-CH_2-\underset{CH_3}{\overset{CH_3}{\underset{|}{C^{\oplus}}}} BF_3OH^{\ominus}$$

Termination

$$CH_3-\underset{CH_3}{\overset{CH_3}{\underset{|}{C}}}\left[CH_2-\underset{CH_3}{\overset{CH_3}{\underset{|}{C}}}\right]_n CH_2-\underset{CH_3}{\overset{CH_3}{\underset{|}{C^{\oplus}}}} BF_3OH^{\ominus} \longrightarrow CH_3-\underset{CH_3}{\overset{CH_3}{\underset{|}{C}}}\left[CH_2-\underset{CH_3}{\overset{CH_3}{\underset{|}{C}}}\right]_n CH_2-\underset{CH_3}{\overset{CH_3}{C}}=CH_2$$

$$+ H^{\oplus} + BF_3OH^{\ominus}$$

Figure 28 Mechanism of cationic chain polymerization of isobutylene

Anionic polymerization causes the formation of so-called "living" polymers because the ionic sites remain active.

Utilization of anionic catalysis in the manufacture of block copolymers is described in Chapter 8.

Olefins with electron-donating groups attached to the double bond can be polymerized by cationic chain polymerization. Examples include vinyl ethers, isobutylene, butylene rubber, tetrahydrofuran, and formaldehyde (see also Table 11). The mechanism of the cationic chain polymerization of isobutylene using boron trifluoride and a trace of water as the catalyst is shown in Figure 28. (Several other proton transfer processes are also possible.)

Polyethylene (PE) $-\!\!\left[\!-CH_2CH_2\!-\!\right]_{\!n}\!-$

Monomer:	Ethylene	
Comonomers:	Butene-1, hexene-1, 4-methylpentene, octene-1	
Polymerization:	LDPE:	Free radical initiated chain polymerization
	ULDPE, LLDPE, HDPE:	Ziegler-Natta or transition metal oxide catalyzed chain polymerization
Major Uses:	ULDPE; LLDPE, LDPE:	Film (60 %), toys and housewares (10 %), wire and cable coverings (5 %)
	HDPE:	Blow-molded items (33 %), injection-molded items (19 %), film (11 %), pipes and fittings (8 %)
Major Producers:	Allied-Signal (Paxon), ATO (Lacqtene) BASF (Lupolen) Bayer (Baylon), BP (Technigram) Chevron (PE) Dow Chemical (LDPE, Dowlex, HDPE), Du Pont Canada (Sclair), Eastman (Tenite), Exxon (Escorene) Hoechst (Hostalen), Huels (Vestolen) ICI (Alkathene), Montedison (Fertene, Moplen), Phillips (Marlex), Quantum (Petrothene), Soltex (Fortiflex), Union Carbide (Unival)	

Polyethylene initially was a homopolymer of ethylene, but today most polyethylenes are copolymers of ethylene with 1 to 10 % of α-olefins, such as 1-butene, 4-methylpentene, 1-hexene, and 1-octene. LDPE and some HDPE grades are ethylene homopolymers. LDPE is a branched polymer with only 40 to 60 % crystallinity. The other polyethylene grades are predominantly linear, with small amounts of branching caused by the comonomer. The linear polyethylenes are produced using Ziegler-Natta catalysts, or transition metal catalysts (chromium or titanium with alkylaluminum cocatalysts). The densities of polyethylene range from 0.880 to 0.957 g/cm^3 as shwon below in Table 13.

In 1968 Phillips Petroleum Company pioneered low pressure, linear low density polyethylene (LLDPE). The LLDPE polymers are linear but have a significant number of branches introduced by using comonomers, such as butene-1 or octene-1. The comonomer content is about 8 to 10 % at a density of 0.920 g/cm^3, and the melting point is about 15 °C higher than in the corresponding LDPE. The linearity provides strength, while the branching provides toughness. The modulus and ultimate tensile properties are significantly improved over branched LDPE. A 1.5 mil film of LLDPE has the same strength as a 5 mil branched LDPE film. LLDPE has penetrated all the nonclarity film markets, except for the shrink film. It is also used for grocery bags, heavy-duty shipping sacks, diaper liners, and agricultural films.

LLDPE is produced by slurry polymerization (Phillips), solution polymerization (Dow Chemical, Du Pont of Canada, Mitsui), and gas phase polymerization using a fluid bed (Union Carbide's Unipol and Himont's Spherilene processes) or a stirred bed (Amoco). Recently Mobil has developed a new catalyst for the Unipol process to produce high strength hexene copolymer LLDPE. The global demand for LLDPE in 1990 was 41 billion pounds.

Ultra-low-density polyethylene (ULDPE) is also a copolymer or terpolymer of ethylene with α-olefins or a combination of α-olefins. Longer chain comonomers, such as 1-octene, provide improvements in tear strength and toughness. ULDPE is a softer, lower modulus polymer, which provides better sealability. It is used in applications requiring superior toughness and puncture resistance, flexibility, and sealability. Specific film end-use applications include food packaging, shrink film, heavy-duty film, and heat-seal layers. Coextrusion and blending with HDPE and PP to improve tear and impact strength is another major application.

The homopolymer, LDPE, is manufactured under high pressure (15,000 to 50,000 psi) and temperature (to 350 °C) using peroxide initiators. If the reaction is conducted in an autoclave, a polymer with narrow molecular weight distribution is obtained. Production in a tubular reactor yields polymers with broader molecular weight distribution. The amorphous polymer has significant amounts of long-chain branching, and a melting range of 107 to 120 °C. LDPE has high impact strength, low brittleness temperature, flexibility, film transparency, and outstanding electrical properties. It was used during World War II to insulate radar cables. Today it is still used for power and communication cable insulation. Major markets are now in food packaging products, industrial sheeting, and trash bags. Photodegradable polyethylene for some food packaging products is made by using a small portion of carbon monoxide as a comonomer. Such copolymers undergo degradation upon exposure to sunlight and moisture. Eventually such a polymer is converted by aerobic digestion into carbon dioxide and water. Anaerobic digestion would produce methane.

High density (low pressure) polyethylene is produced either by Ziegler-Natta catalysis or by metal oxide catalysis, usually referred to as Phillips catalysis. The Ziegler-Natta catalysts are a mixture of titanium chloride and chlorodiethylaluminum with aluminum/titanium ratios between 1:1 and 2:1. The polymerization is usually conducted well below the melting point of the polymer, using hydrocarbon solvents as diluents. The insoluble polymer precipitates. Typical reaction temperatures are 50 to 70 °C, and the reaction time ranges from 1 to 4 hours. The solid supported, chromium oxide catalyzed reaction (Phillips process) is run continuously in a hydrocarbon solvent at 125 to 160 °C. Comonomers, such as α-olefins, are used to the extent of up to 1 % in HDPE production.

Third generation polyolefin catalysts, which produce highly regular polymers with narrow molecular weight ranges, were developed recently by Exxon, Dow Chemical, and Mitsui. These single-site constrained geometry metallocene catalysts are an extension of the Ziegler-Natta chemistry. Exxon and Mitsui combined their efforts in single-site catalyst and gas phase process technologies to produce low density polyethylene polymers having lower melting points, higher transparency and greatly reduced amounts of hexane soluble impurities. These new polymers are intended for film and wire and cable applications. The new catalysts also have the capability to copolymerize styrenic and vinyl monomers.

The melting range of the more linear HDPE is 130 to 138 °C. High molecular weight (HMW) HDPE has molecular weights of 200,000 to 500,000, and ultrahigh molecular weight (UHMW) HDPE has molecular weights of 3 million to 6 million. The UHMW-HDPE polymers have no melting points, and they are processed by compression molding or ram extrusion. Applications include bulk material handling, profile extrusion, chemical pump parts, and snow plow edges. Recently, high strength fibers were produced from UHMW-HDPE.

HDPE is produced in molecular weights ranging from 10,000 to several million, it has high crystallinity, low to medium stiffness and hardness, medium to extremely high toughness, no reduction in toughness to – 40 °C, and unrestricted usage in contact with food. Major applications include blow molding applications (milk bottles, household chemical bottles, cans, drums, containers, fuel tanks), corrugated pipe, and wire and cable insulation. HMW-HDPE is used in electrical, film, pressure pipe, large part blow molding, and extruded sheet applications. The global demand for HDPE in 1990 was 27 billion pounds.

For the polymerization of ethylene it is essential to produce the monomer in high purity. The polymerization catalysts for LDPE are added to the monomer in concentration of less than 0.005 % by weight, and the gaseous monomer is compressed prior to heating in the reactor. LDPE is produced by three processes: low pressure; high pressure, tubular reactor; and high pressure, stirred autoclave. Ethylene under the reaction conditions is an incompressible liquid, and therefore reaction occurs in ethylene solution. Excess ethylene is recycled, and no purification is required. The molten polymer is extruded and granulated.

Polyethylene is a partially amorphous and partially crystalline polymer. Side chain branching is the key property factor in controlling the degree of crystallinity. HDPE has up to 90 % crystallinity, while LDPE exhibits crystallinity as low as 50 %. Increasing the density increases stiffness, tensile strength, hardness, heat and chemical resistance, opacity, and barrier properties, but reduces impact strength and stress-crack resistance. A comparison of PE materials is shown in Table 13.

Table 13 Types of Polyethylene

PE	Density (g/cm³)	Crystallinity	Melting Point (°C)
ULDPE	0.880~0.915	Low	
LLDPE	0.918~0.940	High	130
LDPE	0.910~0.955	40~60 %	107~120
HDPE	0.941~0.954	High	130~138
HMW-HDPE	0.944~0.954	High	
UHMW-HDPE	0.955~0.957	High	~*

* Has no melting point.

The trend in polyethylene production is toward the manufacture of more cost-effective products with improved physical properties. Therefore, the consumption of LLDPE will surpass that of LDPE in the years to come. Also HDPE consumption will increase because of growth of the high molecular weight (HMW and UHMW) polymers, which have superior properties over the regular HDPE polymers.

A variety of copolymers of ethylene are manufactured using a free radical initiated chain polymerization technology similar to LDPE production. The major copolymers are listed in Table 14.

Table 14 Ethylene Copolymers

Comonomer	Name	Copolymer Content (%)
Vinyl acetate	EVA	5~50*
Vinyl alcohol	EVOH	27–48
Methyl acrylate	EMA	20–40
Ethyl acrylate	EEA	15–30
Acrylic acid	EAA	3–20

* If vinyl acetate content exceeds 50 %, the copolymer is named VAE.

The most widely used comonomer content in EVA is 5 to 20 %. The addition of the comonomer reduces the crystallinity of polyethylene, which improves clarity, low temperature flexibility, impact strength, and crack resistance. Melting and heat-seal temperatures of the copolymers are reduced. EVA copolymers are more permeable to gases and water vapors. Major applications include flexible packaging, shrink wrap, heavy-duty shipping sacks, produce bags, bumper pads, flexible toys, tubing, and wire and cable. Major producers are USI (Ultrathene), Du Pont (Elvax), and Exxon (Escorene). The U. S. consumption of the EVA copolymers exceeds 1 billion pounds per year.

The EVOH copolymers are made by the hydrolysis of EVA copolymers. They are hydrophilic polymers that absorb moisture. EVOH copolymers are often used as barrier structures in rigid and flexible packaging. Major producers are EVAL Company, Kuraray, Nippon Gohsei, and Solvay.

EMA copolymers are used for the manufacture of flexible gloves, in medicinal packaging, and as blending components in polyolefins to improve impact resistance, heat-seal response, and toughness. The major producer is Chevron. Also EEA copolymers are used as blending components in LDPE and LLDPE. Other applications include special hose and tubing, films, disposable gloves, balloons, diaper liners, hospital sheeting, and hot-melt adhesives.

Ethylene copolymers with acrylic and methacrylic acid are used to improve adhesion. The metal salts of ethylene/methacrylic acid copolymers are sold as ionomers by Du Pont (Surlyn). Most commercial ionomers are based on sodium or zinc salts. Ionomers have outstanding puncture and low temperature impact resistance, and they are used in sporting goods (golf ball covers) and automotive (bumper pads and guards) and footwear applications. Polyketones can be obtained by an alternating copolymerization of ethylene and carbon monoxide using homogenous palladium catalysts.

Low chlorination transforms polyethylene from a thermoplastic material into an elastomer. The low chlorine products (22 to 26 % chlorine) are softer, more rubberlike, more soluble, and more compatible than the original polyethylene. Random substitution reduces chain order and thereby also the crystallinity. A typical commercial product is Dow Chemical's Tyrin. Crosslinking by free radical initiated vulcanization is also used in some formulations, especially for cable jacketing. An acrylonitrile-chlorinated polyethylene-styrene terpolymer (ACS) is produced in Japan. When polyethylene is chlorinated in the presence of sulfur dioxide, a product is obtained with chlorosulfonyl groups in the chain. This elastomer can be vulcanized with metal oxides. Du Pont sells this polymer under the trade name Hypalon. Hypalon is used in automotive applications, for wire and cable insulation, and as pond liners. The longest elongation is obtained with LDPE containing about 27 % chlorine and 1.5 % sulfur.

Polyvinyl fluoride (PVF)

$$-\left[\begin{array}{c}CH_2CH\\|\\F\end{array}\right]_n-$$

Monomer:	Vinyl fluoride
Polymerization:	Free radical initiated chain polymerization
Major Use:	Films
Major Producers:	Du Pont (Tedlar), Dynamit Nobel (Dyflor)

Polyvinyl fluoride is a tough and transparent high melting crystalline polymer, which retains useful properties over a temperature range of -70 to $110\,°C$. PVF is converted into thin films by plasticized melt extrusion, and it is used in coatings for aluminum and galvanized steel for exterior residential and industrial building siding, and for aircraft interior panels.

Polyvinyl chloride (PVC)

$$\left[-CH_2CH- \atop | \atop Cl \right]_n$$

Monomer: Vinyl chloride (ethylene and chlorine or acetylene and HCl)
Polymerization: Free radical initiated chain polymerization
Major Uses: Pipe and fittings (51 %), films and sheets (11 %), flooring materials (10 %), vinyl siding (7 %), wire and cable insulation (7 %), automotive parts (5 %), adhesives and coatings (5 %)
Major Producers: ATO (Lacqvyl), BASF (Vinoflex, Vinuran), Dexter (Alpha), Dynamit Nobel (Trosiplast), Goodrich (Geon), Huels (Vestolit), ICI (Corvic, Welvic), Montedison (Sicron), Occidental (Oxyblend), Shintech, Wacker (Vinnol)

Polyvinyl chloride is produced by free radical initiated chain polymerization. In the United States, about 82 % of the PVC is manufactured by suspension polymerization, 10 % by bulk polymerization, and 8 % by emulsion polymerization. Solution polymerization is also employed to some extent. The vinyl chloride monomer traditionally has been manufactured from acetylene and hydrogen chloride, but availability of inexpensive ethylene made the chlorination route more attractive. A combination of direct chlorination and oxychlorination of ethylene provides a balanced process, because the hydrogen chloride generated in the pyrolysis of ethylene dichloride is used in the oxychlorination of ethylene.

$$CH_2 = CH_2 + Cl_2 \longrightarrow ClCH_2CH_2Cl$$

$$CH_2 = CH_2 + HCl + 1/2\, O_2 \longrightarrow ClCH_2CH_2Cl + H_2O$$

$$ClCH_2CH_2Cl \xrightarrow{\Delta} CH_2 = CHCl + HCl$$

The control of vinyl chloride monomer escaping into the atmosphere during handling in the production of PVC has become important because cases of angiosarcoma, a rare type of liver cancer, were found among workers exposed to the monomer. Manufacturing processes have been modified to prevent exposure to vinyl chloride.

PVC is the most versatile of all plastics because of its blending capability with plasticizers, stabilizers, and other additives. It is used as a rigid homo- or copolymer, a flexible PVC containing a plasticizer, as plastisol (dispersion) or as a latex. The homopolymers are formulated with thermal stabilizers for protection against thermal depolymerization. The crystalline (syndiotactic) homopolymer has a T_g of 81 to 82 °C, and a density of 1.38 to 1.4 g/cm³. About 90 % of the PVC produced is used in the form of a homopolymer. The global demand for PVC in 1990 was 40 billion pounds.

Copolymers with 5 to 20 % of vinyl acetate provide improved toughness and processing. For example, a copolymer containing 13 % vinyl acetate is used in vinyl floor tiles. Copolymers containing vinylidene chloride or acrylonitrile are manufactured to increase thermal stability and improve solubility. These copolymers are used in coating applications. Copolymers with propylene afford clear and tough polymers. Specialty terpolymers of vinyl chloride with vinyl acetate and ethylene are used as an emulsion in coating applications. PVC has the advantage over other thermoplastic polyolefins of built-in fire retardancy because it is 57 % chlorine. Since PVC is readily chlorinated, polymers with a chlorine content of up to 70 % (CPVC) are manufactured. Blends of PVC with chlorinated ethylene, polyurethane elastomers, or butadiene polymers are formulated to impart improved impact properties.

Approximately 60 % of PVC is used in building and construction, mainly in rigid pipes, flooring, and vinyl siding. Plasticized PVC is used mainly in wire and cable insulation and in packaging. Flexible calendered film and sheeting is used in many diverse applications ranging from pool liners and roof coatings to vinyl-coated fabrics. PVC plastisols (dispersion resins) are used in flooring, artificial leather, wall coverings, and carpet backing. Other important applications include injection molding of pipe fittings, electrical outlets boxes, and parts for automotive bumpers.

Polyvinylidene fluoride (PVDF) $\ \ \ -\!\!\left[CH_2CF_2\right]_n\!\!-$

Monomer: Vinylidene fluoride
Polymerization: Free radical initiated emulsion and suspension polymerization
Major Uses: Bottles and pipes
Major Producers: ATO (Foraflon), Dynamit Nobel (Dyflor), Kureha (KF), Pennwalt (Kynar), Solvay (Solef)

PVDF is a crystalline, high molecular weight polymer. Vinylidene fluoride adds predominantly to the growing chain in head-to-tail sequence, but 4 to 6 % of the monomer units add head-to-head and tail-to-tail, depending on the polymerization process. The melting points of PVDF range from 154 to 184 °C, its upper use temperature is 150 °C, and the density is 1.78 g/cm^3.

PVDF is used as an electrical insulator material in computer back-panel wire, jacketing for aircraft hookup wire, and in geophysical cables. In the chemical processing industry it is used in piping systems, lined tanks, and in valves and pumps for corrosive fluid handling.

Polyvinylidene chloride (VDC) $-\!\!\left[\!-CH_2CCl_2-\!\right]_{\!n}\!\!-$

Monomer: Vinylidene chloride (dehydrochlorination of trichloroethane)
Polymerization: Free radical initiated chain polymerization
Major Uses: Film and sheeting for food packaging
Major Producers: Dow Chemical (Saran)

The vinylidene chloride monomer is produced commercially by dehydrochlorination of trichloroethane with lime or caustic. It can be homo- or copolymerized by emulsion or suspension processes. Emulsion polymers are used directly as latexes or after coagulation and dewatering in coatings. Suspension polymers are used for melt processing. The melting point of the homopolymer is 198 to 205 °C, but decomposition of the homopolymer occurs rapidly at 210 °C, which makes processing difficult. Therefore copolymers with vinyl chloride (13 %), acrylates, or acrylonitrile melt lower and make processing easier.

VDC homo- and copolymers have low permeability to gases, liquids, flavor, and aroma. Therefore, they are used in food packaging, as household wraps, and in pharmaceuticals and cosmetics packaging. Multilayer films containing a barrier layer of VDC are used for packaging fresh red meat, processed meat, cheese, and poultry. The coating resins are used for paper, fabric, and container liner applications.

Polychlorotrifluoroethylene (PCTFE) $-\!\!\left[\!-CClFCF_2-\!\right]_{\!n}\!\!-$

Monomer: Chlorotrifluoroethylene
Polymerization: Free radical initiated chain polymerization
Major Uses: Electrical/electronics, medical instruments, wire and cable
Major Producers: ATO (Voltalef), Daikin, 3M (KEL-F)

Polychlorotrifluoroethylene (PCTFE) is a high melting (211 to 216 °C), high molecular weight, crystalline polymer with an upper use temperature of 250 °C. The high fluorine content protects against attack by chemicals and oxidizing agents. PCTFE has a density of 2.1 g/cm^3. PCTFE films have the lowest water vapor transmission of any transparent plastic film. This polymer is also used in chemical processing equipment, and in cryogenic and electrical applications.

Also a trichlorofluoroethylene copolymer with ethylene (ECTFE) is produced by Ausimont (Halar). This copolymer has a density of 1.68 g/cm^3 and a melting point of 240 °C, but its upper use temperature is only 165 °C.

This copolymer is used in mass transit cables, in fire alarm cables, in valve and pump components, and in connectors. ECTFE coatings and linings protect metals from corrosive environments. Glass fiber backed ECTFE sheets are used as tank liners.

Polytetrafluoroethylene (PTFE) $-\!\!\left[\mathrm{CF_2CF_2}\right]_n\!\!-$

Monomer:	Tetrafluoroethylene (chloroform and HF)
Polymerization:	Free radical initiated suspension and emulsion polymerization
Major Uses:	Gaskets, seals, coatings for chemical process equipment, wire and cable, electrical components, nonadherend surfaces
Major Producers:	ATO (Soreflon), Ausimont (Halon, Algoflon), Daikin (Floraflon), Du Pont (Teflon), Hoechst (Hostaflon), ICI (Fluon)

Polytetrafluoroethylene was discovered at Du Pont in 1947. PTFE is manufactured by free radical initiated suspension polymerization to give a granular product, or by emulsion polymerization, which produces a fine powder. The high molecular weight white, translucent polymer has a density of 2.14 to 2.20 g/cm^3; it has excellent heat and chemical resistance, but it cannot be processed as a thermoplastic. The homopolymer has a crystallite melting point of 327 °C, and it is stable up to 280 °C.

The monomer is commercially produced from chloroform and hydrogen fluoride as shown:

$$\mathrm{CHCl_3 + HF \longrightarrow CHClF_2 + 2HCl}$$

$$\mathrm{2CHClF_2 \xrightarrow[600\,°C]{\Delta} CF_2{=}CF_2 + 2HCl}$$

PTFE has outstanding resistance to chemical attack, and it is insoluble in all organic solvents. Its impact strength is high, but its tensile strength, wear resistance, and creep resistance are low in comparison to other engineering plastics. Because of its high melt viscosity, it cannot be processed by conventional molding techniques. Molding powders are processed by melt and sinter methods used in powder metallurgy. Ram extrusion is possible.

Major applications of PTFE include liners and components for chemical processing equipment, high temperature wire and cable insulation, and molded electrical components. Reinforced PTFE applications include bushings and seals in compressor hydraulic applications, and pipe liners. As fillers or reinforcements, 15 to 25 % glass fibers, 15 % graphite, and 60 % bronze have been used. The fine powder PTFE is used in wire coatings, thin-walled tubing, and tapes. No-stick surfaces for home cookware are perhaps the best known applications for PTFE.

To produce perfluorinated polymers that can be processed by injection molding and extrusion, copolymers of tetrafluoroethylene with hexafluoropropylene (FEP copolymers) and with perfluoroalkylvinyl ethers (Teflon PFA) are produced (see tetrafluoroethylene copolymers).

7 Polyolefins

Polypropylene (PP)

$$-[CH_2CH(CH_3)]_n-$$

Monomer:	Propylene
Polymerization:	Ziegler-Natta catalyzed chain polymerization
Major Uses:	Fiber products (30 %), automotive parts (15 %), packaging (15 %), toys and housewares (5 %), appliance parts (5 %)
Major Producers:	Amoco, ATO (Lacqtene P), BASF (Novolene), Eastman (Tenite), Exxon (Escorene), Fina (Fina), Himont (Profax), Hoechst (Hostalen PP), Huels (Vestolen P), ICI (Propathene), Montedison (Moplen), Rhône-Poulenc (Napryl), Shell, Soltex (Fortilene), USI (Norchem NPP)

Isotactic polypropylene is a stereospecific polymer in which the propylene units are attached in a head-to-tail fashion and the methyl groups are aligned on the same side of the polymer chain. The isotactic polypropylene cannot crystallize in the same fashion as polyethylene (planar zigzag), since steric hindrance by the methyl groups prohibits this conformation. Isotactic polypropylene crystallizes in a helical form in which there are three monomer units per turn of the helix. PP is the lightest of the major plastics, with a specific gravity of 0.90 to 0.91 g/cm^3 and a melting range of 165 to 170 °C.

Polypropylene is made entirely by low pressure processes, using Ziegler-Natta catalysts (aluminum alkyls and titanium halides). Usually 90 % or more of the polymer is in the isotactic form. Production processes include solvent (slurry) polymerization, gas phase polymerization, liquid monomer process, and the Montedison-Mitsui high yield process. About 60 % of the worldwide polypropylene capacity is based on Himont's Spheripol Process. This process allows production of polymer particles with soft cores surrounded by a solid shell. In this manner softer and more flexible polypropylene can be obtained.

Major markets for the homopolymers are filaments and fibers, automotive and appliance components, housewares, packaging containers, furniture, and toys. Films are used as pressure-sensitive tapes, packaging films, retortable pouches, and shrink films. Polypropylene fibers are produced by an oriented extrusion process. Major advantages are their inertness to water and microorganisms and low cost.

About 20 % of PP is manufactured as copolymers (2 to 10 % ethylene). Terminal block impact copolymers of propylenen with ethylene are made in the reactor or by compounding PP with ethylene-propylene rubbers. Random copolymers (1.5 to 3.5 % ethylene) are produced to improve clarity and flexibility by breaking up the crystallinity of PP. They also have a lower melting point. Random copolymers are used in blow molding and film applications. Unoriented films are used in packaging of consumer products, such as shirts, hosiery, bread, and produce. Oriented films provide high clarity and gloss, and they are used in shrink wrap applications. Heat-seal films are used in food packaging. Molded products include video cassette cases, storage trays, and toys.

Poly(1-butene)

$$-[CH_2-CH(CH_2CH_3)]_n-$$

Monomer:	1-Butene
Polymerization:	Ziegler-Natta catalysis
Major Uses:	Pipes, cable insulation
Major Producers:	Akzo (Tetrafil), Huels (Vestolen BT), Shell (Duraflex)

Poly(1-butene) is made by polymerization of 1-butene using Ziegler-Natta catalysis in excess monomer. The isotactic polymer has a high molecular weight (230,000 to 750,000) and a crystallinity of about 50 %. Its density is 0.915 g/cm^3, and the melting point is 125 to 130 °C.

The polymer crystallizes from the melt in a tetragonal crystalline modification, which slowly transforms irreversibly into a stable twined hexagonal crystalline modification. Also copolymers of ethylene and 1-butene are manufactured, which have lower melting points (90 to 118 °C) depending on the ethylene content.

The homopolymer has excellent creep resistance and retention of properties up to 80 °C, and it is compatible with polypropylene. Major applications include pipe, cable jacketing, and food and meat packaging. The ethylene copolymer is used in hot-melt adhesives and sealants.

Poly(4-methylpentene)

$$-[CH_2-CH(CH_2CH(CH_3)_2)]_n-$$

Monomer:	4-Methylpentene (dimerization of propylene)
Polymerization:	Ziegler-Natta catalysis
Major Uses:	Transparent medical and laboratory equipment, films for food packaging
Major Producers:	Akzo (Methafil), Mitsui (TPX), Phillips (Crystalor)

The polymerization of 4-methylpentene, produced by dimerization of propylene, is carried out in hydrocarbon solvent or in excess monomer, using Ziegler-Natta type catalysts. The polymer is a stiff isotactic thermoplast with good electrical properties. It has a density of 0.83 g/cm^3 and a melting point of 235 to 240 °C, and it can be used from 0 to 122 °C. Some of the commercial polymers are copolymers to enhance the optical and mechanical properties.

Applications include laboratory and medical ware, vending machines, appliances, dairy equipment, lighting, and electrical/electronic equipment. Packaging uses include cook-in-containers for hot air and microwave ovens, and coated paperboard for take-out foods and bakery products.

Polystyrene (PS)

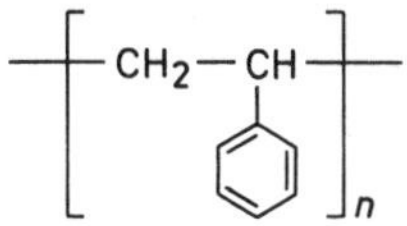

$$\left[-CH_2-CH(C_6H_5)- \right]_n$$

Monomer:	Styrene
Polymerization:	Free radical initiated chain polymerization
Major Uses:	Extruded items (46 %), molded items (34 %), expandable beads (13 %)
Major Producers:	Amoco, Arco (Dylene), ATO (Lacqrene), BASF (Polystyrol, Styropor), Chevron, Dow Chemical, Hoechst (Hostyren), Huels (Vestyron), Huntsman, Montedison (Edistir)

The polystyrene homopolymer is made by bulk or suspension polymerization of styrene monomer. This product is referred to as crystal PS. The homopolymer has a density of 1.05 g/cm³. Impact polystyrene is made by feeding a mixture of styrene monomer and elastomer into a series of reactors, conducting the free radical polymerization of styrene using heat or catalysts to about 90 % conversion of the monomer. Residual monomer is removed by distillation and the final product is pelletized. Expandable polystyrene is made from crystal PS using physical or chemical blowing agents.

Styrene monomer is produced from benzene and ethylene. Another process to produce styrene monomer and propylene oxide simultaneously was introduced by Halcon, now Arco, in 1969. Both processes are summarized in Figure 29.

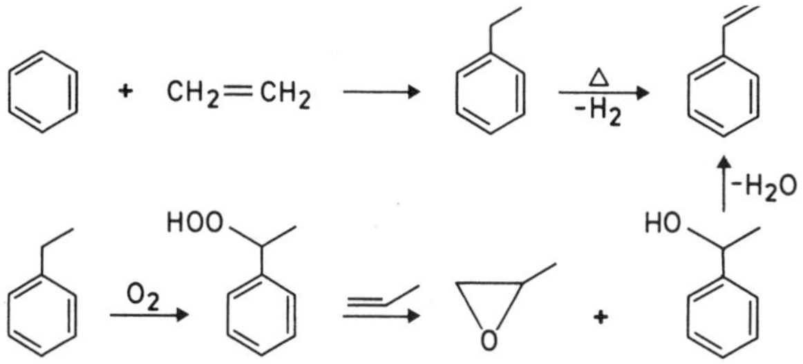

Figure 29 Manufacture of styrene monomer

Crystal PS is an amorphous homopolymer of styrene that offers stiffness, dimensional stability, and good optical properties. However, the UV-stability of crystal PS is poor. Applications of crystal PS include injection-molded products (household wares, video cassettes, appliances, business machine housings and furniture); extruded items include sheets used in thermoformed packages; and biaxially oriented film used in blister packs. Also expandable PS (EPS) is made by foam molding of crystal PS.

Impact polystyrene (IPS, HIPS) is produced commercially by dispersing small particles of butadiene rubber in styrene monomer. This is followed by mass prepolymerization of styrene and completion of the polymerization either in mass or in aqueous suspension. During prepolymerization, styrene starts to polymerize by itself, forming droplets of polystyrene with phase separation. When nearly equal phase volumes are obtained, phase inversion occurs, and the droplets of polystyrene become the continuous phase in which the rubber particles are dispersed. The impact strength increases with rubber particle size and concentration, while gloss and rigidity are decreasing. The stereochemistry of the polybutadiene has a significant influence on properties and a 36 % *cis*-1,4-polybutadiene provides optimal properties. The major application of impact styrene is also packaging. Examples include fast food packaging items, such as cups, lids, and take-out containers, and containers for food, fruit juices, and dairy products.

Expandable polystyrene (EPS) is made using pentane or other low boiling inert solvents as blowing agents. The foam is usually prepared by "steam-chest" molding. A polystyrene foam with a density of 1 lb/ft^3 has 97 % of its volume made up by air. Styrofoam is used because of its resistance to heat flow, as well as its energy absorption, buoyancy, high stiffness to weight ratio, and low cost per volume. It is used as insulation material in the building and construction industry, in disposable containers, and in protective packaging. The global demand for polystyrene in 1990 was 20 billion pounds.

Other aromatic polyolefins are made from p-methylstyrene, p-t-butylstyrene and divinylbenzene.

Polyvinyl pyridine

$$-[CH_2-CH(C_5H_4N)]_n-$$

Monomer:	Vinyl pyridine
Polymerization:	Free radical initiated chain polymerization
Major Uses:	Adhesives, rubber goods
Major Producer:	Reilly Industries

2-Vinyl pyridine can be homopolymerized by free radical or anionic polymerization. However, the major use for this compound is in copolymerization with other olefin monomers, especially in rubbers and fibers. In the latter the pyridine acts as a site for dye fixation.

The best-known products are vinyl pyridine-butadiene-styrene terpolymers, which are composed of 70 % butadiene, 15 to 21 % styrene, and 9 to 15 % vinyl pyridine. They are used as adhesives to bond textile fibers to natural and synthetic rubbers in the manufacture of tires and mechanical rubber goods. The latexes are manufactured by a batch or continuous emulsion polymerization process. They usually contain 40.5 % polymer solids in water.

Polybutadiene (Butadiene Rubber, BR) $-\!\!\left[CH_2CH\!=\!CHCH_2\right]_{\!n}\!\!-$

Monomer:	Butadiene (by-product in steam cracking, dehydrogenation of butylenes or butane from refineries and natural gas)
Polymerization:	Ziegler-Natta catalyzed chain polymerization
Major Uses:	Tire treads, carcass, and sidewall (91 %); belts, hoses, gaskets, and seals
Major Producers:	Anic, Bayer, Bunawerke Huels, Bridgestone/Firestone Tire & Rubber Company, Goodyear Tire & Rubber Company, Michelin

Polybutadiene (butadiene rubber: BR) is a stereospecific (mainly 1,4-*cis*) elastomer made by solution polymerization of butadiene using Ziegler-Natta catalysts. Slight changes in catalyst compositon produced drastic changes in polymer composition, as shown in Table 15.

Table 15 Effect of Catalysts on Polybutadiene Composition

Catalyst	% *trans*-1,4	% *cis*-1,4	% 1,2
Et_3Al/VCl_3	97-98	–	2-3
$Et_3Al/TiCl_4$	49	49	2
$Et_3Al/Ti(OBu)_4$	0-10	–	90
$Et_3Al/Cr (acac)_3$	18	12	70
$Et_2AlCl/CoCl_2$	3	93-94	3-4

The stereochemistry of polybutadiene is important if the product is to be used as a base polymer for further grafting. A polybutadiene with 60% *trans-1,4*-, 20% *cis-1,4*-, and 20 % 1,2-configuration is used in the manufacture of ABS. The low temperature impact strength is related to the glass transition temperature of the components. The T_g of *cis*-1,4-polybutadiene is –108 °C, while the trans product has a T_g of –14 °C.

Butadiene rubber is usually blended with natural rubber or styrene-butadiene rubber (SBR) to improve tire tread performance, especially wear resistance. Polybutadiene is also used in the manufacture of impact-modified polystyrene products. The terms "rubber" and "elastomer" are used interchangeably, usually referring to a material that can be stretched to at least twice its original length and returns quickly to approximately its original length after release of the force applied in the stretching. Relatively low cost materials with good resiliency and durability are used in major tire and tread applications. Specialty elastomers are higher priced and used in specific applications (Table 16 lists the presently used commercial elastomers).

Table 16 Commercial Elastomer Products

Type	Major Applications	Properties
Natural rubber	Tires, bushings and couplings, seals, footwear, belting	Low hysteresis, excellent fatigue resistance, excellent physical properties but fair chemical resistance.
Styrene-butadiene (SBR)	Tires, tire products, footwear, wire and cable, adhesives	Good physical properties when reinforced with carbon black, but poor resistance to sunlight and ozone and chemicals.
Polybutadiene	Tires, hose, and belts	Good low-temperature properties and adhesion to metals, poor resistance to chemicals.
Polyisoprene	Tires and tire products, belting, footwear, flooring	Similar to natural rubber, but excellent flow characteristic during molding.
Ethylene-propylene	Nontire automotive products, hose, wire, and cable	Outstanding resistance to oxygen and ozone, poor tire cord adhesion, poor fatigue resistance.
Butyl	Tire inner tubes, inner liners, seals, coated fabrics	Outstanding retention of properties, but low resiliency and poor resistance to oils and fuels.
Nitrile	Seals, gaskets, footwear, hose	Excellent resistance to oils and solvents, poor low-temperature flexibility and poor resistance to weathering.
Chloroprene	Mechanical automotive goods, wire and cable, hose, footwear	High strength but difficult to process.
Silicone	Wire and cable insulation, encapsulation of electronic components	Outstanding electrical and high temperature properties, poor tear and abrasion resistance.
Polyurethane	Fibers, industrial tires, mechanical goods, footwear, wire and cable	High strength, abrasion and oil resistance, relatively high price.
Chlorosulfonated polyethylene	Wire and cable, hose, footwear, pond liners	Good abrasion resistance, but only moderate oil resistance.
Epichlorohydrin	Seals, gaskets, wire and cable	Good chemical resistance, but poor processability.
Polysulfide	Adhesive, sealants, binders, hose	Outstanding resistance to oil and solvents, but low strength
Fluorocarbons	Aerospace applications, high quality seals and gaskets	Outstanding heat resistance, highest priced elastomer

Low volume specialty polybutadiene products include liquid *trans*-1,4-polybutadienes used in protective coatings inside metal cans, low molecular weight liquid 1,2 polybutadiene (60 to 80 % 1,2-content) used as potting compounds for transformer and submersible electric motors and pumps, and liquid hydroxyterminated polybutadiene resins used in polyurethane elastomers. Global demand for polybutadiene in 1990 was 1,176,000 tons.

7 Polyolefins

Polyisoprene

$$\left[-CH_2-\underset{\underset{CH_3}{|}}{C}=CH-CH_2- \right]_n$$

Monomer:	Isoprene (dehydrogenation of pentenes; dimerization of propylene; isobutylene and formaldehyde)
Polymerization:	Ziegler-Natta catalyzed chain polymerization
Major Uses:	Passenger car tires (60 %), footwear, mechanical goods, sporting goods, sealants, and caulking compounds
Major Producers:	Anic, CPS, Goodyear Tire & Rubber Company, Shell Nederland

Polyisoprene is produced by stereospecific solvent polymerization, using Ziegler-Natta type catalysts. In this manner 98 to 99 % *cis*-1,4-polyisoprene is obtained. Anionic polymerization of isoprene, using organolithium catalysts, also produces a polymer high in *cis*-1,4 content. The *cis*-1,4-polyisoprene is chemically identical to natural rubber but is cleaner, lighter in color, more uniform, and less expensive to process. Using mixed VCl_3-$TiCl_3$-R_3Al catalysts, iosprene polymerizes to *trans*-1,4-polyisoprene, which is similar to natural Balata rubber.

Copolymers of isoprene with butadiene are also readily obtainable because both monomers have the same reactivity. For example, using $Al(i-Bu)_3/TiCl_4$ as the catalyst, both monomers enter into the copolymer chain in the same configurations (98 % cis for isoprene, 70 % cis for butadiene) as in the homopolymers.

Isoprene monomer (2-methyl-1,3-butadiene) is not readily available from oil cracking processes, and therefore several routes are used for its synthesis. Shell Chemical Company and B. F. Goodrich Chemical Company extract isoamylene fractions from catalytically cracked gasoline streams and produce isoprene by subsequent catalytic dehydrogenation. The reaction is as follows:

$$CH_3-\underset{\underset{CH_3}{|}}{C}=CH-CH_3 \;+\; CH_2=\underset{\underset{CH_3}{|}}{C}-CH_2CH_3 \;\xrightarrow{-H_2}\; \text{(isoprene)}$$

Dimerization of propylene is another viable route to isoprene. Several steps are involved: dimerization of propylene to 2-methyl-1-pentene, isomerization to 2-methyl-2-pentene, and pyrolysis to isoprene and methane. Goodyear operates a 60,000 ton/year isoprene plant in the United States that is based on the dimerization of propylene. Another process based on isobutylene and formaldehyde is used in Western Europe, the USSR, and Japan.

$$\text{(isobutylene)} \;+\; 2\,CH_2O \;\longrightarrow\; \text{(dioxane)} \;\longrightarrow\; \text{(isoprene)} \;+\; CH_2O \;+\; H_2O$$

In Italy, Enichem produces isoprene from acetone and acetylene.

Polyisoprene elastomers are used in the construction of passenger car tire carcasses and inner liners, and also in the manufacture of heavyduty truck and bus tire treads. The increasing demand for radial-ply tires in the United States is likely to increase the use of isoprenic elastomers (blends) for passenger car and truck tires. Mechanical goods, footwear, sporting goods, sealants, and caulking compounds are other important applications for polyisoprene elastomers.

Polychloroprene

$$\left[-CH_2-\underset{Cl}{C}=CH-CH_2- \right]_n$$

Monomer:	Chloroprene (dimerization of acetylene and hydrogen chloride)
Polymerization:	Free radical initiated chain polymerization
Major Uses:	Conveyor belts, automotive parts, seals, bridge mounts, expansion joints, wire and cable jacketing, roofing membranes
Major Producers:	Bayer, Distugil, Du Pont (Neoprene)

Polychloroprene was the first commercially successful synthetic elastomer. It was invented by the Carothers group and introduced by Du Pont in 1932 under the trade name Neoprene. Polychloroprene is manufactured by free radical initiated emulsion polymerization, and it contains primarily linear *trans*-2-chloro-2-butenylene units arising from 1,4-addition polymerization. In polychloroprene, a total of 10 stereoregular polymers are envisioned because of the asymmetry caused by the chloro group. The possible stereoregular polymers derived from chloroprene are shown in Table 17.

The chloroprene monomer is manufactured from acetylene according to the following reactions:

$$2\,HC\equiv CH \xrightarrow{CuCl} H_2C=CH-C\equiv CH \xrightarrow{HCl} \text{(2-chloro-1,3-butadiene)}$$

Compounding of neoprene rubber is very similar to that of natural rubber. Vulcanization is achieved with a combination of zinc and magnesium oxide and added accelerators and antioxidants. Carbon black or mineral fillers are sometimes added also. Adhesive grade polychloroprene often contains other compatible resins.

Because of its overall durability, neoprene rubber is used chiefly where a combination of deteriorating effects exist. Industrial rubber goods include uses such as conveyor belts, diaphragms, hose, seals, and gaskets. Automotive uses include hose, V-belts, and weather stripping. Some major uses in construction are for highway joint seals, pipe gaskets, and bridge mounts and expansion joints.

Neoprene latex has the same general properties as natural rubber latex and can be used in the same applications. Some of the applications of neoprene latexes are found in adhesives, asphalt modification, cement modification, coatings, and sealants. The total global consumption of polychloroprene in 1990 was 248,000 metric tons.

7 Polyolefins

Table 17 Polychloroprene Structures

Type of Polymer	Structure
1,2-Isotactic 1,2-Syndiotactic	$\left[-CH_2-\underset{\underset{\parallel}{C}}{\overset{Cl}{C}}-\right]_n$ (with =CH$_2$)
3,4-Isotactic 3,4-Syndiotactic	$\left[-CH_2-CH-\right]_n$ with side group =C(Cl)
1,4-*Trans* head-to-tail	$-CH_2\diagdown C=C \diagup^{H} \diagdown CH_2-CH_2 \diagup^{Cl} C=C \diagup^{CH_2-} \diagdown H$; Cl / \\
Head-to-head combinations *trans* form	$-CH_2\diagdown C=C \diagup^{Cl} \diagdown CH_2-CH_2 \diagup^{Cl} C=C \diagup^{CH_2-} \diagdown H$; H /
Tail-to-tail combination *trans* form	$-CH_2\diagdown C=C \diagup^{H} \diagdown CH_2-CH_2 \diagup^{H} C=C \diagup^{CH_2-} \diagdown Cl$; Cl /
1,4-*Cis* head-to-tail	$-CH_2\diagdown C=C \diagup^{CH_2-CH_2} \diagdown H \quad \diagup^{CH_2-} C=C \diagdown H$; Cl / , Cl /
Head-to-head combination *cis* form	$-CH_2\diagdown C=C \diagup^{CH_2-CH_2} \diagdown Cl \quad \diagup^{CH_2-} C=C \diagdown H$; H / , Cl /
Tail-to-tail combination *cis* form	$-CH_2\diagdown C=C \diagup^{CH_2-CH_2} \diagdown H \quad \diagup^{CH_2-} C=C \diagdown Cl$; Cl / , H /

8 Olefin Copolymers

Introduction

There are four types of copolymer. The most commonly encountered are alternating or random copolymers, in which the two monomer units alternate in random statistical distribution in the polymer chain. The other types are block copolymers, in which like polymer units occur in relatively long sequences; graft copolymers, in which different monomer units are grafted onto a linear polymer chain, and star block copolymers (see Figure 30).

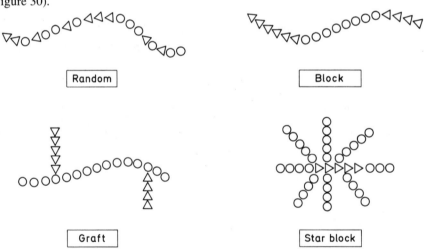

Figure 30 Types of copolymers

Copolymers are created to improve the properties of the homopolymers, most commonly the mechanical properties, such as impact resistance, or to incorporate other desirable properties, such as fire retardancy, dyeability, or solvent and chemical resistance. Copolymerization processes are similar to homopolymerization processes. For example, free radical copolymerization is not very selective, and copolymer composition varies in relation to comonomer composition. In contrast, ionic copolymerization is more selective, and the copolymer composition is relatively unaffected by the comonomer composition. The most common copolymers and terpolymers are shown in Table 18.

8 Olefin Copolymers

Table 18   Some Common Industrial Copolymers and Terpolymers

Molding Compounds	
SAN	Styrene-acrylonitrile
ABS	Acrylonitrile, butadiene, styrene
Ionomers	Ethylene-methacrylic acid salts
Elastomers	
SBR	Styrene, butadiene
NBR	Acrylonitrile, butadiene
EPDM	Ethylene-propylene-diene monomer
Butyl rubber	Isoprene-isobutylene
Thermoplastic olefins	Blends of EPM or EPDM with PP or PE
Fluoroelastomers	Vinylidene fluoride copolymers

In the construction of block copolymers, the selective reactivities of the monomers have to be considered. Block copolymers based on butadiene and styrene can be produced by solvent, emulsion, suspension, or bulk polymerization, because the diene reacts preferentially, leaving the styrene to terminate the reaction. If a difunctional initiator is used, the styrene can be added at the end to give a styrene-butadiene-styrene triblock. Three-stage processes are also feasible. In this case a monofunctional initiator, such as alkyllithium, is used, followed by sequential monomer additions. Coupling of diblocks is another viable alternative (see the section on thermoplastic olefin elastomers in this chapter).

The grafting reactions can be achieved by abstracting a labile hydrogen from the polymer chain, using a free radical initiator. Termination of the side chain occurs by disproportionation or radical combination. Of course, radical combination leads to crosslinking. The degree of crosslinking can be minimized by low grafting.

Blending or alloying different polymers is another way to combine polymers to create materials with superior properties (see Chapter 9).

Styrene-Acrylonitrile Copolymers (SAN)

$$\left[CH_2CHCH_2CH \atop \quad\; |\qquad\quad\; | \atop \quad\; CN\quad\;\; Ph \right]_n$$

Monomers:	Acrylonitrile, styrene
Polymerization:	Free radical initiated chain polymerization
Major Uses:	Appliances, housewares, packaging, automotive, electronics
Major Producers:	ATO (Lacqsan), BASF (Luran), LNP (Thermocomp), Monsanto (Lustran SAN)

Styrene-acrylonitrile copolymers are random amorphous linear copolymers. They possess transparency, high heat deflection properties, and excellent gloss and chemical resistance. Their properties are controlled by acrylonitrile content and molecular weight.

Increases in acrylonitrile content and molecular weight generally result in property improvements. However, processability will suffer. Special grades include UV-stabilized, barrier, weatherable, antistatic, and glass-reinforced materials. Also olefin-modified SAN materials are made by polymerizing styrene and acrylonitrile onto an olefinic elastomer.

Because of their thermal stability, SAN copolymers are used in molding "dishwasher-safe" housewares such as refrigerator meat and vegetable drawers, blender bowls, vacuum cleaner parts, humidifier parts, and detergent dispensers. Also glass-reinforced dashboard components and battery cases are moldet from SAN. Over 35 % of the total SAN production is used in the manufacture of ABS blends.

Acrylonitrile-Butadiene-Styrene Terpolymers (ABS)

$$\left[-CH_2\underset{CN}{CHCH_2}CH=CHCH_2CH_2\underset{\bigcirc}{CH} - \right]_n$$

Monomers:	Acrylonitrile, butadiene, styrene
Polymerization:	Free radical initiated chain polymerization
Major Uses:	Appliance and automotive, pipe and fittings, telephone and business machine housings
Major Producers:	ATO (Lacqran), BASF (Terluran), Bayer (Novodur), Dow Chemical (Magnum), G. E. (Cycolac), Monsanto (Lustran ABS), Montedison (Urtal)

The ABS terpolymers are a family of products formed basically from three different monomers. The overall balance of properties will vary depending upon monomer ratios, molecular weight, and additives. Acrylonitrile contributes heat resistance, high strength, and chemical resistance; butadiene contributes impact strength, toughness, and low temperature property retention; and styrene contributes gloss, processability, and rigidity. For some special grades of ABS a fourth monomer is used. For example, heat-resistant grades incorporate α-methylstyrene, transparent grades incorporate methyl methacrylate, and maleic anhydride is used in grades intended for alloying with polycarbonates. Flame retardancy can be achieved using halogenated additives or by alloying with PVC or CPE.

Most ABS is made by grafting styrene and acrylonitrile directly upon a BR or SBR rubber latex in a batch or continuous emulsion polymerization process. Also high rubber content ABS is sometimes blended with SAN made by either emulsion, suspension, or bulk polymerization. ABS terpolymers are processable by all techniques commonly used with thermoplastics. As in the case of SAN, drying is required prior to use in molding machines.

Specialty grades include electroplating grades (automotive grilles and exterior decorative trim), high temperature resistant grades (automotive instrument panels, power tool housings), fire-retardant grades (appliance housings, business machine housings, television cabinets), and structural foam grades (high strength/weight ratio in molded parts). Major applications are in refrigerator doors and trays, business machine and computer housings, pipe and fittings, and automotive trim. Blow-molded ABS is used in seating, interior and exterior doors, and luggage.

Ethylene-Methacrylic Acid Copolymers (Ionomers)

$$\left[-CH_2CH_2CH_2\underset{COO^{\ominus}}{\overset{CH_3}{\underset{|}{\overset{|}{C}}}}- \right]_n$$

Monomers: Ethylene, methacrylic acid
Polymerization: Free radical initiated chain polymerization
Major Uses: Packaging, automotive parts, recreational footwear
Major Producer: Du Pont (Surlyn)

Ionomer is a generic term for polymers containing interchain ionic bonding. Du Pont, which introduced ionomers in 1964, remains the only supplier of these polymers. Surlyn copolymers are sodium or zinc salts of copolymers derived from ethylene and methacrylic acid. The properties of ionomers vary according to type and amount of cation, molecular weight, and composition of the base copolymer. The ionized carboxyl groups create ionic crosslinks in the intermolecular structure. These interchain forces produce properties normally associated with a crosslinked structure. However, ionomers are thermoplastic materials that can be processed in conventional molding machines.

Ionomers are transparent and are resistant to abrasion and solvents. The combination of these properties offers potential applications where toughness and see-through clarity are desired. Ionomers also have excellent electrical properties.

Ionic elastomers have also been obtained from ethylene-propylene-diene terpolymer (EPDM) (discussed later in this chapter). This material contains about 1 mole % sulfonate groups appended to some of the unsaturated groups of EPDM. The sulfonic acid groups are neutralized with zinc acetate to form the zinc salt of the EPDM elastomer. The effect of the physical crosslinks provided by the ionic groups is demonstrated by the softening behavior of the sulfonated EPDM as shown in the modulus versus temperature curve in Figure 31. Physical association of the ionic groups provide pseudo-crosslinks, and the ionic groups from different molecules aggregate to form domain areas as shown in Figure 32. Addition of a foaming agent gives rise to formation of thermoformable flexible foams, which are useful for weather-stripping and shoe soles.

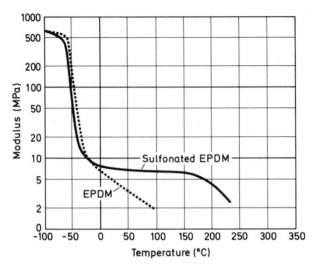

Figure 31 Modulus versus temperature of EPDM and sulfonated EPDM

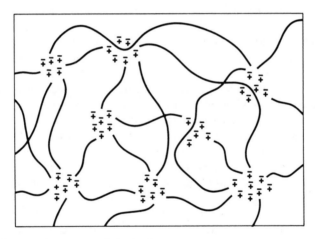

Figure 32 Domain structures in ionic elastomers

Styrene-Butadiene Rubber (SBR)

$$\left[-CH_2CH=CHCH_2CH_2CH(C_6H_5) - \right]_n$$

Monomers:	Styrene, butadiene
Polymerization:	Free radical initiated chain polymerization (mostly emulsion polymerization)
Major Uses:	Tires and rubber thread (70 %), mechanical goods (20 %), latex (10 %)
Major Producers:	Bayer, Enichem, Firestone Tire & Rubber Co., General Tire & Rubber Co., Huels, B. F. Goodrich, Goodyear Tire & Rubber Co., Phillips, Shell, Uniroyal

The random styrene-butadiene copolymer (SBR) is the most widely used synthetic rubber. Global consumption in 1991 was about 2,711,000 tons. Most SBR made by emulsion polymerization contains 23 to 25 % styrene randomly dispersed with butadiene in the polymer chain. The stereochemistry of the butadiene portion is 14 to 19 % *cis*-1,4-, 60 to 69 % *trans*-1,4-, and 17 to 21 % 1,2-configuration. Typical emulsifiers used in the production of SBR are fatty acid soaps (sodium stearate), and the reaction is initiated with potassium peroxydisulfate (hot rubber process) or *p*-menthanehydroperoxide (cold rubber process). The emulsifiers commonly used in the manufacture of olefin polymers are shown in Table 19.

Table 19 Emulsifiers and Initiators Used in the Emulsion Polymerization of Olefin Monomers

Monomer	Emulsifier*	Initiator**
Acrylonitrile	ai	HP, OP, PS
Butadiene	fs, rs	PS
Chloroprene	aas, fs, rs, ci	PS
Isoprene	fs, rs	PS
Styrene	aas, fs, rs, ci	HP, OP, PS
Vinyl chloride	as, fs	HP, OP, PS
Vinylidene chloride	as	PS

* as = Alkylsulfates or sulfonates, aas = alkylarylsulfonates,
ai = anionic emulsifiers, ci = cationic emulsifiers,
fs = fatty acid soaps, rs = rosin acid soaps.
** HP = Hydrogen peroxide, OP = organic peroxide,
PS = peroxydisulfate.

In the emulsion polymerization of SBR, long-chain mercaptans are used as chain transfer agents. The chain transfer agents serve to terminate growing polymer chains and to initiate new chains.

RSH + M$_n$ ⟶ M$_n$H + RS·

RS· + M ⟶ RSM·

The concentration of chain transfer agent controls the molecular weight distribution of the copolymer. Table 20 shows the composition of the diene structure in the hot and the cold rubber processes.

Table 20   Polydiene Composition in SBR Rubbers

Process	Temperature (°C)	% *trans*-1,4	% *cis*-1,4	% 1,2
Hot rubber process	50	62	14	23
Cold rubber process	5	72	7	21

Dry SBR (also called solid rubber) is about 14 to 25 % styrene and 75 to 85 % butadiene, while SBR latexes are emulsions of styrene-butadiene copolymers containing also about 23 to 25 % styrene. The latexes are used for the manufacture of foam backing for carpets and for adhesives and molded foam applications. The solid rubber products are used mainly in tires and tire products. Sometimes a solution polymerization process is used for the manufacture of SBR. The solution process provides polymers with more linear configuration and with better control over molecular weight distribution.

The styrene-butadiene elastomers have gained their dominant position in the tire industry because of their excellent abrasion resistance and their cost/performance/processing balance relative to polybutadiene rubber, natural rubber, and polyisoprene rubber. The total synthetic rubber production in the world in 1991 was 10.4 million metric tons, while the total natural rubber produced was 5.3 million metric tons.

Styrene-Butadiene Block Copolymers

Monomers:	Styrene, butadiene, isoprene
Polymerization:	Anionic block copolymerization
Major Uses:	Footwear, automotive parts, hot-melt adhesives
Major Producer:	Phillips (K-Resin), BASF (Styrolux), Shell (Kraton)

Conventional rubbers are crosslinked by primary valence bonding, whereas thermoplastic elastomers are crosslinked by secondary valence bonding, such as van der Waals interaction, dipole interactions, hydrogen bonding, or ionic bonding. The secondary valence crosslinking is reversible at elevated temperature or under the influence of solvents. In a broad sense, all chemically uncrosslinked polymers can behave as thermoplastic elastomers. Starting from the glassy state, amorphous thermoplastics show rubber elasticity and then viscous flow with increasing temperature, but usually thermoplastic elastomers are segmented copolymers of butadiene and isoprene with styrene. They have

8 Olefin Copolymers

a diene central segment (block), which is terminated with polystyrene segments. The center elastomer segment can also consist of ethylene-butylene rubber, which provides better ozone resistance because of absence of unsaturation.

The block copolymers can be produced with difunctional initiators (sodium complex of naphthalene, dilithium compounds), by first polymerizing the diene followed by polymerization of styrene or by three-stage processes using alkyllithium compounds as initiators: that is, using sequential monomer additions.

The difunctional initiation of the polymerization of styrene using the sodium complex of naphthalene can be illustrated as follows:

$$[\text{naphthalene}]^{\ominus} \; Na^{\oplus} + PhCH=CH_2 \rightleftharpoons \text{naphthalene} + Ph\dot{C}H\overset{\ominus}{C}H_2 \; Na^{\oplus}$$

$$2\; Ph\dot{C}H\overset{\ominus}{C}H_2 \; Na^{\oplus} \longrightarrow Na^{\oplus}\;\overset{\ominus}{C}HCH_2CH_2\overset{\ominus}{C}H\;Na^{\oplus} \xrightarrow{PhCH=CH_2} Na^{\oplus}\;\overset{\ominus}{C}HCH_2[CH_2CH]_n CH_2\overset{\ominus}{C}H\;Na^{\oplus}$$

with Ph substituents on the indicated carbons.

Ph = –⟨phenyl⟩

Dienes are preferentially polymerized in the presence of styrene, leaving styrene for the end to produce a styrene-butadiene-styrene triblock polymer. The polystyrene blocks have a molecular weight of 10,000 to 20,000, whereas the polydiene block has a molecular weight of 40,000 to 100,000. The styrene-butadiene block copolymers usually contain more styrene than butadiene.

Star-shaped block copolymers are prepared by adding divinylbenzene to a styrene-butadiene diblock copolymer that still contains lithium atoms at its ends. The star-shaped block copolymers were developed primarily for adhesives, but they may also be used in combination with high impact polystyrene or ABS.

Another method involves the copolymerization of the diene and styrene in solution with an anionic catalyst under conditions that cause sequential monomer addition. This two-block copolymer is then reacted with a coupling agent to form a chain of double length, terminated on both ends with polystyrene blocks:

$$2\,SSSS\ldots BBB\,Li + COCl_2 \longrightarrow SSSS\ldots BBB-CO-BBB\ldots SSSS + 2\,LiCl$$

Styrene-butadiene block copolymers are transparent and high impact polymers used in footwear applications, such as tennis shoes and sneakers; in injection molding of unit soles for lower priced men's and women's shoes; in hot-melt adhesive applications (pressure-sensitive tapes); and in molding and extrusion of automotive parts. The styrene-butadiene block copolymers can be blended with other thermoplastics (PS, SAN, or styrene-methyl methacrylate copolymers) to improve economics and to enhance physical properties. If clarity is not essential, blending with PP, HIPS, PS or PC is also possible. The global consuumption of styrenic block copolymers in 1991 was 295,000 tons.

Block copolymers of hydrogenated styrene with ethylene and butylene, or ethylene and propylene are sold by Shell under the Kraton G tradename.

Nitrile Rubber (NBR)

$$\left[\begin{array}{c} CH_2 CHCH_2 CH=CHCH_2 \\ | \\ CN \end{array}\right]_n$$

Monomers:	Acrylonitrile, butadiene
Polymerization:	Free radical initiated chain polymerization (mostly emulsion polymerization)
Major Uses:	Footwear, belts, hoses, gaskets adhesives, sealants
Major Producer:	Bayer, BP, Enichem, B. F. Goodrich (Hycar), Goodyear Tire & Rubber Co. (Chemigum), Uniroyal (Paracryl)

Nitrile rubber (acrylonitrile-butadiene copolymer) is a unique specialty elastomer, which has been manufactured in the United States since 1939. It is unique in its excellent resistance to oil over a wide range of temperatures. The oil resistance increases with increasing amounts of acrylonitrile in the copolymer, while the low temperature properties and resilience improve with decreasing acrylonitrile content. NBR is also noted for its high strength and excellent resistance to abrasion, water, alcohols, and heat. Disadvantages are poor dielectric properties and poor resistance to ozone.

Almost all NBR is produced by the free radical initiated emulsion polymerization of butadiene and acrylonitrile, and, like SBR production, both "hot" and "cold" types are available. The acrylonitrile content ranges from 18 to 50 % by weight. Since the reaction ratios are not unity, the copolymer composition changes with conversion. The butadiene portion is more in the trans configuration, which gives better tensile properties. Applications include footwear, belts and belting, hoses, gaskets, adhesives, sealants, coatings, and wire and cable insulation. Global consumption in 1991 was 250,000 tons.

Ethylene-Propylene Elastomers

$$\left[\begin{array}{c} CH_2 CH_2 CH_2 CH \\ | \\ CH_3 \end{array}\right]_n$$

Monomers:	Ethylene, propylene
Polymerization:	Ziegler-Natta catalyzed chain polymerization
Major Uses:	Automotive parts, radiator and heater hoses, seals, single-ply roofing, pond liners, wire and cable
Major Producers:	DSM, Du Pont (Nordel), Enichem, Exxon (Vistalon, Exxral), Genesis (Adpro), Himont (Hifax), Huels (Vestoprene), Monsanto (Santoprene, Vyram), Polysar (Epcar), Uniroyal (Royalene)

Ethylene-propylene elastomers are synthetic polymers with outstanding resistance to oxygen, ozone, and heat. They were introduced in the United States in 1962. Two types of ethylene-propylene elastomer are currently being produced: ethylene-propylene copolymers (EPM) and ethylene-propylene-diene terpolymers (EPDM). The EPM copolymers are saturated, and they require vulcanization with free radical generators. In

8 Olefin Copolymers   77

contrast, EPDM elastomers are produced by polymerizing ehtylene and propylene with a small amount (3 to 9 %) of nonconjugated diolefins. The pendant side chain permits vulcanization with sulfur. The density of EPM is 0.86 to 0.87 g/cm^3, and the glass transition temperature is –60 °C.

Ethylene-propylene copolymers are produced by solution polymerization using Ziegler-Natta type catalysts. Copolymers containing 1 to 1.5 ethylene units for every propylene unit are most desirable, because long blocks produce undesirable crystallinity. EPDM terpolymers are produced similarly by adding 3 to 9 % of 1,4-hexadiene or ethylidene norbornene to the monoolefin mixture. For the copolymerization of ethylene and propylene gas phase technologies are available from Amoco/Chisso; Union Carbide/Shell and BASF/ICI/Norchem. Union Carbide/Shell is using a fluid bed catalysts. The principle control in these processes is the gas composition within the reactor.

Many thermoplastic olefin (TPO) elasotmers are blends of EPM or EPDM with PP or PE (see Polyolefin Alloys and Blends in Chapter 9). In these blends both phases are totally dispersed. Three categories of TPOs are produced: physical blends of PP and rubber, thermoplastic vulcanizates dispersed in PP (e. g. Monsanto's Santoprene and Vyram' Himont's Hifax and Hüls's Vestoprene), and reactor-modified TPOs. In the latter category the elastomers are reacted in situ with propylene. The recyclability of TPO's has helped them to displace RIM molded polyurethane automotive bumpers, particularly in Europe.

TPOs are mainly used in nontire automotive applications, including radiator and heater hoses, body and chassis parts, bumpers, weather-strips, seals, and mats. Advantages include low specific gravity and low cost ($ 0.85 to 0.95/lb). The use of EPDM for the manufacture of regular passenger and truck tires has been disappointing, mainly because of poor tire cord adhesion and poor compatibility with most other elastomers. Other applications include appliance parts, wire and cable insulation, impact modification of polypropylene and other thermoplastics, hoses, gaskets and seals, and coated fabrics. EPM is also used as viscosity index additive in lubrication oils. Global consumption of TPOs in 1991 was 603,000 tons.

Butyl Rubber

$$\left[-CH_2-\underset{\underset{CH_3}{|}}{\overset{\overset{CH_3}{|}}{C}}-CH_2CH=\underset{}{C}-CH_2- \right]_n$$

Monomers:	Isobutylene, isoprene
Polymerization:	Cationic chain copolymerization of isobutylene with 0.5 to 2.5 mole % of isoprene
Major Uses:	Tire inner tubes and inner liners of tubeless tires, inflatable sporting goods
Major Producers:	Cities Service Co., Esso, Exxon, Polysar

Butyl rubber was introduced in the United States in 1942. The high level of impermeability to common gases over a wide temperature range and low glass transition temperature

(–70 °C) are the main advantages of butyl rubber. A major disadvantage is its lack of compatibility with SBR, polybutadiene, and natural rubber.

Butyl rubber is produced by a slurry copolymerization process using isobutylene and isoprene (0.5 to 2.5 mole %) in methylene chloride at –100 °C in the presence of aluminum chloride as the catalyst. The isoprene is mainly linked by 1,4-addition.

Tetrafluoroethylene Copolymers

Monomers:	Tetrafluoroethylene, hexafluoropropylene, perfluoroalkylvinyl ether, ethylene
Polymerization:	Free radical initiated suspension, emulsion, and solvent polymerizations
Major Uses:	Electrical applications, membranes, wire and cable, pipes and fittings, conveyor belts
Major Producers:	ETFE: Asahi (Aflon COP), Aussimont (Halon), Du Pont (Tefzel), Hoechst (Hostaflon ET), ICI (FP-FF)
	FEP: Daikin (Neoflon), Du Pont (Teflon FEP)
	PFA: Du Pont (Teflon PFA), ICI (FP-PF)

The tetrafluoroethylene copolymers were invented to obtain melt processable Teflon-like polymers. The comonomers are used to break up the crystallinity in polytetrafluoroethylene. Perfluoroolefin comonomers, such as hexafluoropropylene and perfluoroalkylvinyl ethers achieve thermoprocessability, while maintaining excellent thermal and chemical stability. The use of the lower cost ethylene as comonomer is a compromise with respect to chemical inertness, thermal stability, and physical properties. A 1:1 copolymer of tetrafluoroethylene and ehtylene provides the best balance in properties.

The softening points of ETFE copolymers range from 200 to 300 °C, their densities are 1.7 to 1.77 g/cm^3, and their upper use temperature ranges from 155 to 180 °C, depending on composition. ETFE copolymers are thermoplasts with good thermal stability, chemical inertness, strength, toughness, and nonflammability, and excellent weatherability and electrical properties. Uses include high performance insulation for wire and cable, and clear and tough films.

FEP (fluorinated ethylene-propylene copolymers) are made by copolymerization of tetrafluoroethylene and hexafluoropropylene. Their densities are 2.11 to 2.2 g/cm^3, melting ranges are 253 to 282 °C, the melting points of their crystallites rage from 285 to 295 °C, and their upper use temperature is about 205 °C. FEP copolymers are used in wire and cable insulation, pipes and fittings, electrical applications, and conveyor belts.

The thermoplastic PFA copolymers have a density of 2.1 to 2.2 g/cm^3, the melting point of their crystallites is 300 to 310 °C, and their upper use temperature is 260 °C. PFA copolymers are used for extruded tubings, shapes, and wire and cable jacketing.

8 Olefin Copolymers

Fluoroelastomers

Monomers: Vinylidene fluoride, chlorotrifluoroethylene, tetrafluoroethylene, perfluoropropylene
Polymerization: High pressure free radical aqueous emulsion polymerization
Major Uses: Aerospace industry, heavy-duty industrial equipment, wire and cable jacketing
Major Producers: Asahi (Aflas), Daikin (Dai El), Du Pont (Viton, Kalrez), 3M (Fluorel, Kel-F), Montedison (Tecnoflon)

The three basic types of fluoroelastomer are fluorocarbons, fluorosilicones, and fluoroalkoxyphosphazenes. The fluorocarbons are mainly polyvinylidene fluoride copolymers, such as polyvinylidene fluoride-co-hexafluoropropylene manufactured by Daikin (Dai-El), Du Pont (Viton A), 3M (Fluorel), and Montedison (Tecnoflon). Polyvinylidene fluoride-co-hexafluoropropylene-co-tetrafluoroethylene is manufactured by Daikin (Dai-El G-501) and Du Pont (Viton G). The same terpolymer with peroxide curable sites is manufactured by Du Pont (Viton G). Polyvinylidene fluoride-co-tetrafluoroethylene-co-perfluoromethyl vinyl ether with peroxide curable sites is manufactured by Du Pont (Viton GLT). Polyvinylidene fluoride-co-chlorotrifluoroethylene is manufactured by 3M (Kel-F), polyvinylidene fluoride-co-1-hydropentafluoropropylene and polyvinylidene fluoride-co-tetrafluoroethylene-co-1-hydropentafluoropropylene are manufactured by Montedison (Tecnoflon SL, Tecnoflon-T). Polytetrafluoroethylene-based copolymers are polytetrafluoroethylene-co-perfluoromethyl vinyl ether with peroxide curable sites manufactured by Du Pont (Kalrez) and polytetrafluoroethylene-co-propylene manufactured by Asahi (Aflas 100, 500). In these co- and terpolymers, Cl or CF_3 groups are introduced to break up the crystallinity of polyvinylidene fluoride and polytetrafluoroethylene.

The fluorocarbon elastomers have a fluorine content of 53 to 70 % and a use temperature range from -46 to $316\,°C$. They have low compression sets, high temperature stability, and excellent fuel, oil, and chemical resistance. They are high in price and performance and offer continuous use at about $200\,°C$. The fluoroelastomers with peroxy curable sites can be covulcanized with other vulcanizable rubbers.

Polyfluorosilicones are made from $CF_3CH_2CH_2Si(CH_3)Cl_2$ via a cyclic trisiloxane monomer, which is polymerized with base at elevated temperatures. Examples are Dow Corning's Silastic and FSI from G. E. These polymer have lower temperature stability compared to the fluorocarbons, but they offer flexibility and softness at low temperatures.

The fluoroalkoxyphosphazenes are discussed in Chapter 18.

9 Alloys and Blends

Introduction

The introduction of new polymers is a lengthy and costly process, and very few new polymers were introduced in the 1980s. In contrast, the use of polymer alloys and blends is increasing from year to year. In this manner new marketable polymer entities are introduced rapidly. The use of "modified" thermoplastics accounted for about one-third of all performance plastics in 1990. The possibilities of blending or alloying of the major polymers are infinite, and the marketplace will dictate the tailoring of blends. The principale reason for the blending of polymers is to improve the product/cost performance of a polymer for a specific end-use application. Most pairs of polymers are thermodynamically immiscible, but some polymer blends are compatible and exhibit excellent physical properties that offer advantages over either of the individual polymers. Sometimes a third component is added as a compatibilizer, which allows the blending of two immiscible polymers. For example, blending of polystyrene and polyethylene, two immiscible polymers, can be achieved by adding a graft copolymer containg both PS and PE sections. The graft copolymer acts as a surfactant during the melt-blending operation. Another approach involves so-called interpenetrating networks (IPNs). For example, styrene plus divinylbenzene polymerizes by free radical initiation in the presence of a polyurethane network. IPNs undergo phase separation into microdomains rich in the respective polymer species.

Useful alloys or blends represent a compromise among the properties of the individual polymers. Flame retardancy can often be achieved by blending PVC, CPVC (chlorinated polyvinyl chloride), or CPE (chlorinated polyethylene) with the thermoplastic matrix polymers. Impact resistance can often be improved by blending with rubberlike compounds (EPM, EPDM, BR, NBR, etc.). Sometimes the impact modification can be built into the polymer chain by grafting the monomers to a diene rubber. In some cases the properties of some combinations are synergistic, and the resultant blends or alloys have better properties than either component.

The definition of an alloy or blend is ambiguous. Usually, however, alloys are referred to as polymer mixtures made by mixing the components in the molten stage, and blends are polymer mixtures made by dryblending of the components. Miscible alloys or blends are one-phase polymers with one glass transition temperature. Good mixing is essential in the manufacture of alloys and blends.

Rubber blends are commonly employed in the manufactur of automotive tires (see Chapter 8). Polymer blends are also employed in the manufacture of textile fibers. A commercially important engineering plastic is Noryl, a miscible blend of polyphenylene oxide (PPO) and high impact polystyrene.

Polyblends are made by compounding on mill rolls, in extruders, or in Banbury mixers. Almost all are heterogeneous and consist of a polymeric matrix in which another matrix is embedded. Four types of polyblend can be visualized (see Table 21). Compatibility plays an important part in the design and optimization of polyblends. Higher impact strength is obtained by blending polystyrene either with polybutadiene on which styrene had been grafted, or with a block copolymer of butadiene and styrene. Similarly, a more compatible ABS can be made by blending SAN with polybutadiene on which styrene and acrylonitrile are grafted.

Table 21 Types of Polyblend

Matrix	Dispersed Phase	Improvement	Example
Rigid	Soft	Tougher	Impact PS, ABS
	Rigid	Faster melt flow	PVC, ABS
Soft	Soft	Longer wear	Natural rubber, *cis*-PB
	Rigid	Higher modulus	SBR, PS

Polyblends in which both phases are rigid are called alloys. They are fabricated to improve melt flow and mechanical properties and/or to reduce shrinkage. For example, impact polystyrene is blended into PPO to improve melt flow, while polyblends of PVC with ABS or acrylate graft copolymers have impact strength higher than either of the components. Addition of a rigid polymer to a soft matrix results in an increase in modulus. Such reinforcement also results in an increase in tensile and tear strength.

Polyolefin Alloys and Blends

Thermoplastic olefin elastomers (TPOs) are perhaps the best known polyolefin alloys. They are made by blending LLDPE or PP with EPM or EPDM rubbers. TPOs were introduced by Uniroyal in 1972, and similar blends are produced by Exxon, Goodrich, and Du Pont. They have good mechanical and electrical properties, and high resistance to ozone. A blend of 5 to 20 % EPM or EPDM in PP has improved toughness and low temperature impact strength, a blend of 25 to 45 % EPM or EPDM in PP gives a semi-elastomer, and 50 to 65 % EPM or EPDM in PP affords true elastomers. Himont's Catalloy combines mixtures of polyethylene, polypropylene and polybutene made by a gas phase process. Catalloy can range from low to high modulus polymers.

TPOs are widely used in automotive (air dams and bumper covers), wire and cable, and mechanical goods applications. However, they usually have poor oil and solvent resistance. The low specific gravity in TPOs (0.8 to 1.0 g/cm^3) helps to reduce part weight, and hardness ranges from 70 to 90 shore A are on the market. Major TPO suppliers are Exxon, Goodrich, Du Pont, and Uniroyal in the United States; Bayer (Levaflex EP), DSM (Keltan TP), Hoechst (Hostalen LP), ICI (Propathane TE), and Montedison (Moplen SP) in Europe; and Mitsui (Milastomer) and Sumitomo (Sumitomo TPE) in Japan. Competitors to TPOs are PC/PBT alloys in the major automotive bumper uses. In Europe most of the unpainted bumpers are made from EPDM/PP alloys.

Monsanto's Santoprene blends are also made from EPDM and PP or PE. These elastomers have the capability of thermal crosslinking (in situ vulcanization) during mixing with the thermoplasic matrix. Santoprene blends compete with thermoset rubbers.

LLDPE, LDPE, and HDPE can be blended in a wide range of ratios to produce tailor-made films. Alloys of LLDPE with 20 % or less of LDPE have improved film drawdown. If 50 % of LDPE is used, tensile strength, tear strength, and elongation are improved. Blends of LLDPE with HDPE have lower cost and better puncture resistance at 10 to 30 % HDPE, and better tear strength and heat sealability at 30 to 50 % HDPE. The rapid market acceptance of LLDPE has accelerated the development of LLDPE blends. Typical examples of LLDPE/polyolefin blends are LLDPE/LDPE (50/50) for ice bags, (40/60) for produce bags and merchandise bags, and (30/70) for shirt sacks.

Addition of polyisobutylene (PIB) or butadiene rubber (BR) improves the toughness of polyolefin films. PIB is used at a level of about 30 % in HDPE by Allied in its Paxon rubber modified compounds. Applications of HDPE/PIB alloys include heavy-duty bags, covers and wraps, single-layer films for building and construction, and agricultural uses. LLDPE and its blends compete with HDPE/PIB because they allow a reduction of gauge for heavy-duty bags. Butyl rubber, a copolymer of isobutylene and isoprene, is also used in polyolefin blends. Examples are shown in Table 22.

Table 22 Polyolefin/BR Blends

Blend	Content of BR (%)	Application	Advantage
LDPE/BR	5~7	Thin film	Improved tear strength and barrier properties
HDPE/BR	15–25	Heavy-duty film	Puncture and impact resistance
PP/BR	10–20	Regular film	Improved low temperature impact

Polysar is the principle developer of BR blends.

Also polybutylene (PB) is alloyed with other polyolefins. PB is mainly used in pipe applications because of its heat and creep resistance. Addition of 9 % PP improves the heat sealing characteristics, processability, and tear strength of PB. Also blends of PB with EPDM or EPM (70/30) improve the impact resistance of PB polymers. Hivalloy, produced by Himont, is a polymeric engineering plastic combining olefinic monomers with non-olefinic (Styrene, acrylo-nitrile, maleic anhydride) monomers.

Polystyrene can be modified by the alloying and or grafting of diene rubbers. The toughness of PS is significantly improved, but the transparency is lost. Most high impact polystyrene (HIPS) is produced by grafting styrene monomer to a polybutadiene rubber. Butadiene-styrene copolymers can also be used to modify PS for impact improvement. In this case the clarity of PS is maintained.

Styrene block copolymers consisting of polystyrene end segments and elastomer center segments (Shell's Kraton) are also used as toughness modifiers. Examples are alloys with PS, PP, HDPE, LDPE, and ABS. The use of styrene block copolymers in PP and PE has dropped because of the increased use of EPDM for this purpose.

Ionomer copolymers (polyethylene-*co*-methyl methacrylic acid), produced by Du Pont and by Mitsui in Japan under license from Du Pont, are blended with EPDM or nylon to improve the impact properties.

Polyvinyl chloride Alloys and Blends

PVC is one of the more amenable polymers for alloying or blending, but it cannot be blended with polyethylene because of low miscibility. Adding a third component as a compatibilizer blending becomes possible. For example, a 50/50 blend of PVC/PE can be achieved using about 15 % of EVA as the compatibilizer. PVC/EPDM graft copolymers can be used as masterbatch for alloying with PVC or ABS.

Blends of PVC with acrylates (15 to 20 %), ethylene-vinyl acetate (EVA) copolymers (8 to 10 %), or chlorinated polyethylene (CPE) (2 to 12 %) are widely used to improve processing (hot-melt strength) and impact. For example, small amounts of ethyl acrylate-methyl methacrylate copolymers(MMA/EA), methyl methacrylate-butadiene-styrene (MBS), and methyl methacrylate-acrylonitrile-butadiene-styrene (MABS) are used as PVC modifiers. Applications include pipe fittings, sidings, gutters, films, and blown bottles. PVC/MMA blends are used to improve toughness and thermoformability. Applications are aircraft trays, business machine and computer covers, and mass transit seating.

Also PVC/EVA combinations are used widely. Vinyl chloride monomer can be grafted to EVA up to a 50 % level. Bayer's Levepren is an example. Ethylene-vinyl acetate copolymers, such as Du Pont's Elvaloy and USI's Vyna, can be used in PVC. PVC/EVA competes with PVC/CPE, but it offers better weathering and lower smoke. This is important in the use of PVC in window frames.

CPE (36 % chlorine level) is used in PVC to improve impact. CPE offers in addition good processability and heat stability. A level of 5 % of CPE in rigid PVC achieves the best impact performance. Suppliers of PVC/CPE alloy are Dow Chemical (Dow CPE), Hoechst (Hostalit Z), Osaka Soda (Daisolac), and Showa Denko (Elasten). Typical applications include impact modification for PVC used in roofing, wire and cable, and hoses. CPE is not used in PVC bottles because translucency is required in this application. Acrylate terpolymer elastomers (polyacrylic ester grafted with acrylonitrile and styrene) (AAS) are also used as impact modifiers for PVC.

For rigid applications of PVC, blends of 70 % PVC and 30 % NBR rubber are used, and for elastomeric applications this ratio is reversed. The NBR segment contributes oil and fluid resistance, while the PVC component contributes ozone resistance and appearance. The PVC/NBR blends are used in wire and cable, hose, and belting applications. Major suppliers are B. F. Goodrich (Hycar) and Polysar (Krynac).

PVC is also used to impart flame retardancy to methyl methacrylate polymers. Rohm & Haas (Kydex) and General Tyre (Boltaron) supply 60 to 65 % PVC/40 to 35 % PMMA alloys for sheeting. Major applications are in aircraft seating components, business machine housings, displays, mass transit seating, counter tops, and wall coverings.

ABS and SAN Alloys and Blends

ABS is one of the oldest blend technologies, dating back to the early 1950s. The basic technology of ABS consists of copolymerizing styrene and acrylonitrile in the presence of butadiene rubber. This graft polymer is subsequently blended with a styrene-acrylonitrile copolymer (SAN) to give ABS. Alloying ABS-type polymers with PVC improves their flame retardancy. ABS/PVC alloys range from 25 to 90 % of PVC. The higher the PVC content, the lower the cost. Excellent impact strength is achieved in blends of 75 % PVC and 25 % ABS (Izod impact 22 to 27 ft-lb/in). As the ABS level increases, flame retardancy decreases. ABS with a higher rubber content has the best impact properties. Uses include electrical components, appliances, air conditioner fans and grilles, dishwasher consoles, luggage shells, computer and business machine housings, and housewares. Suppliers of ABS/PVC include G. E., Occidental, Uniroyal, Bayer, DSM, Mitsui, Sumitomo, and Mitsubishi. Also CPE (36 % chlorine level) is used for impact modification and flame retardancy in ABS/SAN and ABS/PVC alloys.

ABS is also blended with polycarbonate (PC). The ABS/PC alloys have improved impact, heat resistance, and processability. Typical applications include typewriter housings, headlight rings, taillight housings, food trays, and appliances. ABS/PC alloys compete with modified PPO, high impact ABS, and ABS/polysulfone alloys. Major suppliers of the ABS/PC alloys include Bayer (Bayblend), Daicel (Cervian), DSM (Ronfaloy), Dow Chemical (Pulse), G. E. (Cycoloy), Mitsubishi (Novomate), Monsanto (Triax 2000), Montedipe (Koblend PCA), and Teijin (Multilon). SAN copolymers with acrylic esters (ASA) are also blended with polycarbonate. The ASA/PC blends have better stress crack resistance and weatherability. Typical applications are parts with outdoor exposure. ASA/MMA blends also have outstanding weatherability and toughness. They are used in spa and hot tub applications.

ABS/nylon alloys are blends of amorphous and crystalline polymers that yield improved chemical resistance and stability at higher temperatures. Typical examples are automotive body panels, connectors, and underhood components.

ABS/polysulfone alloys exhibit outstanding toughness, heat resistance, and good chemical resistance. Uniroyal introduced the Arylon ABS/PSO (57/43) alloys in 1968. They were later sold to Union Carbide and marketed under the trade name Mindel. Typical applications are in the appliance and automotive fields, as well as such items as window handles, faucet components, and food trays. ABS also blends very readily with styrene block copolymers, with acrylic acid esters (to improve weatherability), with ethylene-vinyl acetate copolymers (to improve stress-cracking resistance), and with EPDM (to improve low temperature impact). Also ABS/polyurethane alloys with good impact strength, toughness, and abrasion resistance are produced. An example is Dow Chemical's Prevail.

SAN/EPDM blends were developed by Uniroyal under the trade name Rovel. These alloys were later sold to Dow Chemical, which still markets these products. Similar SAN/EPDM alloys were introduced by BASF, Hitachi, and Toray. Because of their improved weatherability, SAN/EPDM alloys are used in swimming pools steps, pickup truck camper tops, livestock shelters, and other outdoor applications.

Nylon Alloys and Blends

Ionomers and ethylene copolymers are used to improve the impact strength of nylon. In the 1970s Du Pont introduced the tough nylons based on nylon/ionomer (copolymer of ethylene and methyl methacrylic acid) blends. The ionomer level in these alloys is about 15 to 20 %. Other copolymers used in the impact modification of nylons include ethylene-ethyl acrylate, ethylene-vinyl acetate, ethylene-acrylic acid and ethylene-butyl acrylate (at levels of 10 to 25 %). Better impact improvement can be achieved with EPM/EPDM (15 to 25 %). The resultant super tough nylons are not blends, because side chain or linear branching occurs upon mixing in the molten state. The izod impact strength of the general purpose nylon of 1 ft-lb/in. can be improved to 20 ft-lb/in. using this technology. Du Pont is basic in nylon-6,6 and in EPDM. Other suppliers of tough nylon alloys are Bayer and BASF (nylon-6/elastomer), ICI (nylon-6,6/elastomer), Mitsubishi (nylon-6/EPM, EPDM), and Unitika (nylon-6/EPDM). Applications of the ST (supertough) nylons are in all market segments of nylons, such as transportation, industrial, appliances, and recreation. Competitive impact materials are PC/PBT alloys (G. E.'s Valox) and PC/PBT alloys containing acrylic rubbers or EPDM (G. E.'s Xenoy).

Blends of nylon with HDPE are used to improve the resistance of nylon to gasoline. In these blends the PE phase is dispersed in form of lamellar domains throughout the nylon matrix. Such materials are intended for fuel tank appplications and blow-molded bottles for pesticides and agricultural chemicals.

Other nylon alloys are nylon/NBR for hot oil resistance and alloys of nylons with elastomeric polyesteramides for impact improvement. Atochem (Orgalloy), Dexter (Dexlon), ICI (JHB B79), and Sumitomo are marketing nylon-polypropylene alloys. Also blends of nylon-6 and nylon-12 are available from Emser Werke in Switzerland to provide materials with lower moisture absorption. The nylon-11 and 12 polymers have not the impact toughness of nylon-6 and 6,6, therefore graft copolymers of acrylonitrile-methyl acrylate (75/25 %) with nitrile rubbers were developed by Sohio for the use of impact modification of nylon-11 and 12. The impact modifier are used at a level of 6 to 20 %. Nylon/PTFE alloys were also developed to combine a low coefficient of friction with high temperature resistance. Applications for these alloys include bearings, bushings, cams, and door latches.

Polycarbonate Alloys and Blends

Polycarbonate is miscible with a wide variety of polymers. For example, PC is very compatible with PBT, and G. E. has introduced PC/PBT alloys containing acrylate modifiers to improve stress crack resistance, toughness, HDT and processability. These alloys (Xenoy) are intended for automotive bumper and body panel applications. Other PC/PBT blends are Bayer's Makroblend, Mitsubishi's Novarex AM, and Dow Chemical's Sabre. For thickness impact improvement of PC, alloys with olefinic rubbers are used. PC/PE alloys were developed to reduce the notch sensitivity of PC.

PC/PET alloys are also introduced by G.E. (Xenoy) and Polysar (Petsar). These alloys combine PC's toughness, rigidity, and heat distortion with PET's chemical resistance. These alloys are also used for automotive applications where resistance to gasoline, hydraulic fluids, transmission fluids, and motor oil is desired.

Alloys of polycarbonate with polyurethane elastomers have good low temperature impact and chemical resistance. Also PC/PA alloys were developed by Dexter (Dexcarb). Applications again are focused on the automotive industry. Another new family of polycarbonate alloys are PC/styrene-maleic anhydride copolymers. These alloys have properties that fall between PC and ABS. They are used in cookware and food processors and in automotive and medical applications.

Polyester Alloys and Blends

Thermoplastic polyesters are readily alloyable with many other polymers. Property enhancement of PBT can be achieved with blending of ABS, EVA, AS (for moldability), PET (for surface gloss), and EPDM and other rubbers for toughness. Brominated PC can be added for flame retardancy. PBT/PET blends (40 to 60 % PBT) are designed to increase the crystallization rate and lower the cost. In these blends PET bottle scrap can be used to lower the cost even more. Major PBT or PET alloy suppliers are Hoechst-Celanese (Celanex), Du Pont (Rynite), and G. E. (Valox) in the United States, and Atochem, Akzo, Bayer (Pocan PBT), Teijin, and Toray in Europe and Japan. Most blends are glass or mineral filled to increase strength and modulus. Use include engineering applications, such as automotive parts, appliances, electrical/electronic, and industrial applications.

Arco's PBT/styrene-maleic anhydride copolymer blends (Dylark) are used in automotive applications. PBT (PET)/elastomer alloys (60 to 75 % PBT or PET) are designed for impact modification. These alloys are used in automotive and in recreational applications. PET/MMA are designed to improve dimensional stability, and PET/polysulfone alloys, containing high levels of glass or mineral fillers, are also used to improve dimensional stability. The latter alloys are used for electrical connectors, food service devices, and process equipment.

Polyphenylene Oxide Alloys and Blends

One of the most successful commercial alloys is G.E.'s Noryl, a blend of PPO (polyphenylene oxide) and high impact PS. PPO and PS are highly compatible, with glass transition temperatures ranging from 205 °C (PPO) to 100 °C (PS). Elastomer modification through the use of high impact polystyrene provides impact and thoughness. The PPO homopolymer is very brittle and difficult to process. The modified PPOs exhibit good dimensional stability, resistance to moisture, low temperature impact strength, low creep, and good processability. Other suppliers of modified PPO are Asahi and Mitsubishi.

Asahi claims that their materials are made by a melt grafting process, and Mitsubishi uses a copolymer of 2,6-xylenol and 2,3,6-trimethylphenol as the matrix resin. Alloys of PPO and nylon are marketed by G.E. (Noryl GTX).

Styrene block copolymers, such as Shell's Kraton, are also used for impact modification of PPO. Sometimes low levels of PE are added to improve mold release properties. The use of EPDM/styrene monomer grafted polymer improves heat resistance, weathering, and toughness. The solvent and chemical resistance of PPO can be improved by blending with nylon. The combined amorphous and crystalline polymers form two-phase systems with good dimensional stability, modulus, strength, and surface appearance.

End uses for modified PPO include automotive, appliances, electrical/electronic, business machines, industrial fluid handling, recreational, and other applications.

Fluoropolymer Alloys and Blends

Polytetrafluoroethylene is an important commercial alloying polymer that provides in its alloys improved lubricity, fire resistance, and corrosion resistance. Examples of PTFE alloys include PC (5 to 20 % PTFE), polyacetal (5 to 25 % PTFE), PPS (15 to 20 % PTFE), nylon (15 to 20 % PTFE), and polyurethane (15 % PTFE). Glass fibers are often added to the blends to improve mechanical properties.

Almost any peroxide curable elastomer can be blended and covulcanized with fluoroelastomers. Examples include nitrile rubbers, epichlorohydrin polymers, EPDM, and fluorosilicon elastomers (Silastic). The generally used alloys are listed in Table 23.

Table 23   Fluoroelastomer Alloys

Fluoroelastomer	Rubber	Composition (%)	Advantages
Fluorel or Viton G	Nitrile	80/20; 50/50	Upgrades nitrile in thermal resistance
Fluorel or Viton G	EPDM	80/20; 40/60	Combines moisture resistance of EPDM with chemical resistance of the fluorocarbons
Fluorel or Viton G	Epichlorohydrin	60/40; 50/50	Upgrades epichlorohydrin properties
Fluorel or Viton G	Fluorosilicon	60/40; 70/30	Improves low temperature properties

These blends are compounded and processes by conventional rubber techniques, that is, blending, filler addition, roll milling, and master batching where necessary before coworking. They are used in automotive (fuel system seals, hoses, gaskets), industrial (tank seals, expansion joints, and gaskets), and oil field applications (valves, pumps).

10 Acrylics

Introduction

Acrylic polymers are derivatives of acrylic acid, $CH_2 = CHC(O)OR$, where the nature of the R group determines the properties of the polymers. By far the best known applications are acrylic fibers and acrylonitrile copolymers (SAN, ABS, NBR; see Chapter 8). The acrylic polymers discussed in this chapter include polymers and copolymers derived from acrylic and methacrylic acid derivatives, such as nitriles (acrylic fibers), esters, and amides. The polymers derived from acrylic or methacrylic acid are discussed under ionomers (Chapter 8). Acrylic acid and its derivatives are readily available from propylene.

Acrylonitrile is produced by the ammoxidation of propylene. Propane can also be used as feedstock, but the selectivities are only on the order of 30 %. However, the Reppe acetylene reaction is also still being used.

$$CH_2=CH-CH_3 \xrightarrow[NH_3]{O_2} CH_2=CHCN$$

$$CH\equiv CH + HCN \longrightarrow CH_2=CHCN$$

The commercial production of acrylates is also based on propylene. Air oxidation of propylene yields acrylic acid, which is subsequently esterified. The intermediate in this process is acroleine.

$$CH_2=CH-CH_3 \xrightarrow{O_2} [CH_2=CH-CHO] \longrightarrow CH_2=CHCOOH$$

$$CH_2=CHCOOH + ROH \longrightarrow CH_2=CHCOOR$$

The classical process for acrylic esters based on the reaction of ethylene oxide with hydrogen cyanide is still being used in Europe.

$$CH_2-CH_2 + HCN \longrightarrow HOCH_2CH_2CN \xrightarrow[H_3O^{\oplus}]{H_2O, ROH} CH_2=CHCOOR$$
(with epoxide O)

Methacrylic acid esters (methacrylates) are synthesized by air oxidation of isobutylene, or by th acetone-cyanohydrin process, which is similar to the ethylene-cyanohydrin route to acrylates.

10 Acrylics

$$CH_3-\underset{}{C}=CH_2 + O_2 \longrightarrow CH_3-\underset{OH}{\overset{CH_3}{C}}-COOH \xrightarrow{-H_2O} CH_2=\overset{CH_3}{C}-COOH + ROH \longrightarrow CH_2=\overset{CH_3}{C}-COOR$$

$$CH_3-\overset{CH_3}{C}=O + HCN \longrightarrow CH_3-\underset{OH}{\overset{CH_3}{C}}-CN \xrightarrow[H_3O^{\oplus}]{H_2O, ROH} CH_2=\overset{CH_3}{C}-COOR$$

The majority of all commercially produced acrylic polymers are copolymers of an acrylic monomer with one or more different monomers. Radical initiators are used to initiate polymerization. The solution polymerization is used to form soluble polymers. The molecular weight can be controlled by added chain transfer agents, such as thiols. Oxygen has to be excluded because it is an inhibitor of the polymerization reaction. Emulsion polymerization is the most important industrial method for the manufacture of polyacrylates. In this process water, monomers, a water-soluble initiator, and a surfactant are used. Suspension polymerization is a special case of the bulk polymerization. In this process the monomer is suspended in water and a monomer-soluble initiator is used.

Acrylic Fibers

$$-[-CH_2CH(CN)-]_n-$$

Monomer:	Acrylonitrile
Polymerization:	Free radical initiated chain polymerization
Major Uses:	Apparel (70 %), home furnishings (30 %)
Major Producers:	American Cyanamid (Creslan), Anic (Euroacryl), BASF (Zefran), Bayer (Dralon), Cortaulds (Cortelle, Teklan), Du Pont (Orlon), Eastman Kodak (Verel), Monsanto (Acrylan)

The development of acrylic fibers started in the early 1930s in Germany. These polymers were first produced in the United States by Du Pont and Monsanto about 1950.

Acrylic fibers derive their properties from the stiff, rodlike structure of polyacrylonitrile, in which nitrile groups are distributed randomly around the rod surface. Most polymers used in the production of acrylic fibers contain minor amounts of other comonomers. The copolymers are generally produced by aqueous heterogeneous or solution polymerization using batch or continuous processes. Solution polymerization is used to prepare acrylic polymers in a form suitable for wet or dry spinning. Solvents include DMSO, DMF, and aqueous solutions of zinc chloride or various thiocyanates. Comonomers include methyl acrylate, methyl methacrylate, and vinyl acetate. Also small amounts of ionic monomers (sodium styrene sulfonate) are often incorporated to improve dyeability. Modacrylic fibers are composed of 35 to 85 % acrylonitrile and contain comonomers, such as vinyl chloride, to improve fire retardancy.

Acrylonitrile copolymers used for acrylic fibers are white, easy flowing powders with molecular weights of 100,000 to 150,000. The polymers decompose before melting.

Acrylic fibers are used extensively in the apparel market for sweaters, single- and double-knit jersey fabrics, shirts, blouses, hosiery and craft yarns, and pile and fleece goods. In home furnishings they are used for carpets, blankets, curtains, and drapes. Acrylic fibers are more durable than cotton, and they are the best alternative fabric to wool for sweaters.

Acrylic Adhesives

Acrylic adhesives are formulated from functional acrylic monomers, which achieve excellent bonding upon polymerization. Typical examples are the cyanoacrylates developed by Eastman Kodak and ethylene glycol dimethacrylates developed by Borden. Both are one-component systems, and polymerization occurs upon exposure to the atmosphere. Cyanoacrylates are obtained by depolymerization of a condensation polymer derived from a malonic acid derivative and formaldehyde:

$$\begin{array}{c} CN \\ | \\ CH_2 \\ | \\ COOR \end{array} + CH_2O \xrightarrow{-H_2O} \left[\begin{array}{c} CN \\ | \\ C-CH_2 \\ | \\ COOR \end{array} \right]_n \xrightarrow{\Delta} \begin{array}{c} CN \\ | \\ C=CH_2 \\ | \\ COOR \end{array}$$

Cyanoacrylates are marketed as contact adhesives, and they have found numerous applications. Notable uses include surgical glue and dental sealants; morticians use them to seal eyes and lips.

Bisglycol methacrylates are obtained by esterification of ethylene glycol oligomers with methacrylic acid.

$$HO\text{-}[CH_2CH_2O]_n CH_2CH_2OH + 2CH_2=\underset{CH_3}{\overset{|}{C}}-COOH \longrightarrow CH_2=\underset{CH_3}{\overset{|}{C}}-COO\text{-}[CH_2CH_2O]_{n+1}CO-\underset{CH_3}{\overset{|}{C}}=CH_2$$

These materials form highly crosslinked and therefore brittle polymers. Modification with polyurethanes or addition of other polymers, such as low molecular weight vinyl-terminated butadiene-acrylonitrile copolymers or chlorosulfonated polyethylene, have been used to formulate tough adhesives with excellent properties.

The modified dimethacrylate systems can be formulated as two-component adhesives, with a catalyst added just prior tho use or with a polymerization catalyst applied separately to the surface to be bonded. Also, one-component systems have been formulated which can be conveniently cured by ultraviolet irradiation.

Polyacrylates

$$\left[-CH_2-\underset{\underset{COOR}{|}}{CH}- \right]_n$$

Monomers:	Acrylic acid esters
Polymerization:	Free radical initiated chain polymerization
Major Uses:	Coatings, paints, adhesives, fiber modification
Major Producers:	Bayer, Borden, Hoechst-Celanese, Rohm & Haas

Polyacrylates are produced by free radical initiated solvent polymerization. The properties of acrylic ester polymers depend to a large extent on the type of alcohol from which the acrylate is prepared. Solubility in oils and hydrocarbons increases, as expected, with increase in the length of the side chain. Polyacrylates with short side chains are relatively soluble in polar solvents. Usually molecular weights on the order of 100,000 to 200,000 are obtained. Because of their low glass transition temperatures, polyacrylates are permanent plasticizers. The glass transition temperatures of some polyacrylates are listed in Table 24.

Table 24 Glass Transition Temperatures of Polyacrylates

Polymer	T_g (°C)
Methyl acrylate	6
Ethyl acrylate	−24
Propyl acrylate	−45
Isopropyl acrylate	− 3
n-Butyl acrylate	−55
sec-Butyl acrylate	−20
Isobutyl acrylate	−43
tert-Butyl acrylate	43
Cyclohexyl acrylate	16

Polymethyl acrylate is used in fiber modification; polyethyl acrylate is used in fiber modification and in coatings; and polybutyl and poly (2-ethylhexyl acrylate) are used in the formulation of paints and adhesives.

To optimize properties of the polyacrylates, copolymers are often produced. The relative ease of free radical induced copolymerization of a 1:1 mixture of an acrylate ester with other comonomers as shown by the reactivity ratio is given in Table 25. Low values can be offset by adjusting the proportions of the comonomers or the method of introducing them into the polymerization reaction. Values above 25 indicate that good copolymerization is expected. For example, methyl acrylate copolymerizes easily with butadiene and poorly with vinyl chloride. Monomers with functional groups commonly used in copolymerization with acrylates are listed in Table 26.

Table 25 Relative Ease of Copolymer Formation for 1:1 Ratios of Acrylates and Other Comonomers $(R_1/R_2) \times 100$

Comonomer	Methyl acrylate	Ethyl acrylate	Butyl acrylate
Acrylonitrile	53	46	77
Butadiene	66	4.7	8.1
Methyl methacrylate	50.3	30.6	14.6
Styrene	21	16	26
Vinyl chloride	2.7	2.1	1.6
Vinylidene chloride	100	52	55
Vinyl acetate	1.1	0.7	0.6

Table 26 Functional Monomers Used in Copolymerization with Acrylic Monomers

Functional Group	Monomer	Structure
—COOH	Acrylic acid	$CH_2\!=\!CH\!-\!COOH$
	Methacrylic acid	$CH_2\!=\!C(CH_3)\!-\!COOH$
	Itaconic acid	$CH_2\!=\!C(CH_2COOH)\!-\!COOH$
—NH$_2$	Dimethylaminoethyl methacrylate	$CH_2\!=\!C(CH_3)\!-\!COO\!\sim\!N\!<$
—OH	2-Hydroxyethyl acrylate	$CH_2\!=\!CH\!-\!COO\!\sim\!OH$
	N-Hydroxyethyl acrylamide	$CH_2\!=\!CHCONH\!\sim\!OH$
	N-Hydroxymethyl acrylamide	$CH_2\!=\!CHCONHCH_2OH$
—epoxide	Glycidyl methacrylate	$CH_2\!=\!C(CH_3)\!-\!COO\!\sim\!\triangle_O$

Commercially important copolymers include poly (ethylene-co-ethyl acrylate) EEA, and poly (ethylene-co-acrylic acid) EAA, the latter has excellent adhesion to metal and other substrates, and both are used in coatings applications.

Polymethyl methacrylate (PMMA)

$$\left[-CH_2-\underset{\underset{COOCH_3}{|}}{\overset{\overset{CH_3}{|}}{C}}- \right]_n$$

Monomer:	Methyl methacrylate
Polymerization:	Free radical initiated chain polymerization
Major Uses:	Signs, glazing, lighting fixtures, sanitary wares, automotive lenses, solar panels
Major Producers:	Cyro (Acrylite) Du Pont (Lucite), Rohm & Haas (Plexiglas)

PMMA is characterized by crystal clear transparency, unexcelled weatherability, outstanding surface hardness, good chemical resistance, and a useful combination of stiffness, density, and moderate toughness. The heat deflection temperatures range from 74 to 100 °C, with a service temperature of 94 °C. Glass transition temperatures of several polymethacrylates are given in Table 27, and it is evident why PMMA is so often the product of choice. Further improvement of the mechanical properties of PMMA can be achieved by orientation of heat-cast sheets.

Table 27 Glass transition Temperatures of Polymethacrylates

Ester Group	T_g (°C)
Methyl	105
Ethyl	65
n-Butyl	20
n-Decyl	–70
n-Hexadecyl	– 9

Polymethyl methacrylate can be modified by copolymerization of methyl methacrylate with other monomers, such as acrylates, acrylonitrile, styrene, and butadiene. Blending with SBR improves impact resistance. Major applications of PMMA include signs, safety glazing, skylights, aircraft canopies and windows, and lighting fixtures. PMMA molding compounds are used in the automotive industry for taillights and in appliance panels.

The 2-hydroxyethyl ester of methacrylic acid is used as the monomer of choice for the manufacture of soft contact lenses. Copolymerization with ethylene glycol dimethacrylate produces a hydrophilic network polymer, a so-called hydrogel. Hydrogel polymers are glassy and brittle when dry but become soft and plastic after swelling in water. The most important properties of a hydrogel are its equilibrium water content and oxygen permeability.

Polyacrylamide

$$\left[CH_2\underset{CONH_2}{CH} \right]_n$$

Monomer:	Acrylamide
Polymerization:	Free radical initiated chain polymerization
Major Uses:	Flocculant, paper treatment, water treatment, coatings, adhesives
Major Producers:	American Cyanamid, BASF, Dow Chemical, Rhône-Poulenc

Polyacrylamide is a very high molecular weight polymer that exhibits strong hydrogen bonding and water solubility. Polymerization of acrylamide monomer is usually conducted in water, using free radical initiators and chain transfer agents. Copolymerization with other water-soluble monomers proceeds in a similar manner. Polyacrylamide is a linear polymer with the normal head-to-tail structure. A significant amount of chain branching may occur, especially with peroxydisulfate-bisulfite initiators. Often polyacrylamides contain some ionic functionality.

Most of the interest in polyacrylamide is associated with its water solubility. Although originally designed for separation and clarification of liquid-solid phases, the polymers are now also used for binding, thickening, lubrication, and film formation. Major uses include flocculation and settling of aqueous suspensions, especially in the mining industry. Other applications include paper treatment, water treatment, additives in coatings and adhesives, and binder for pigment.

11 Polyvinyl Compounds

Introduction

In polyolefins the R group attached to the olefin monomer is either hydrogen, alkyl, aryl, or halogen. If the R substituent is a cyanide, carboxylic acid ester, or carboxylic acid amide group, the derived polymers are acrylic polymers, as discussed in the preceding chapter. The vinyl polymers are polyolefins in which the R substituents attached to the olefin monomers are bonded through an oxygen atom (vinyl esters, vinyl ethers) or a nitrogen atom (vinyl amides).

As a group the vinyl polymers are an important segment of the plastic industy. They are mainly used in paints and adhesives and in the treatment for textiles and paper. Polyvinylpyrrolidone and polyvinylcarbazole are used in high tech applications, such as photoconductive and electroconductive polymers (see Chapter 19). The commonly used commercial vinyl polymers are listed in Table 28.

Table 28 Commercial Vinyl Polymers

Polymer	Structure of Monomer	Polymerization Initiation	Major End Uses
Polyvinyl acetate	$CH_2=CH-OCOCH_3$	Free radical	Adhesives, paints
Polyvinyl ethers	$CH_2=CH-OR$	Cationic polymerization	Adhesives, coatings
Polyvinylpyrrolidone	$CH_2=CH-N$ (pyrrolidone ring)	Free radical	Cosmetics, textile treatment, plasma volume extender
Polyvinylcarbazole	$CH_2=CH-N$ (carbazole ring)	Ziegler-Natta	Organic photoconductors

The parent compound of the oxygen-bonded derivatives, polyvinyl alcohol, is produced by hydrolysis of polyvinyl acetate (PVA) because vinyl alcohol is not stable in the monomeric form. PVA is also the starting material for polyvinyl butyral (PVB) and polyvinyl formal (PVF). These cyclic six-membered-ring polyacetals are manufactured by reacting PVA with butyraldehyde or formaldehyde in the presence of strong mineral acid. The commercial polymers derived from polyvinyl acetate are listed in Table 29.

Table 29 Commercial Polymers Derived from Polyvinyl Acetate

Polymer	Structure	Major End Uses
Polyvinyl alcohol (PVA)	$-[CH_2-CH(OH)]_n-$	Textile and paper treatment
Polyvinyl butyral (PVB)	$-[CH_2-CH-O-CH(-(CH_2)_2CH_3)-O-CH-CH_2-]_n-$ (6-membered ring with CH bearing $(CH_2)_2CH_3$)	Adhesive
Polyvinyl formal (PVF)	$-[CH_2-CH-O-CH_2-O-CH-CH_2-]_n-$ (6-membered ring with CH_2)	Enamel for wire coating and fibers

The vinyl monomers are usually manufactured from acetylene and the corresponding acids, alcohols, amines, and amides. However, the largest volume product, vinyl acetate, is produced mainly from ethylene.

Polyvinyl Acetate (PVA) $-[CH_2-CH(OCOCH_3)]_n-$

Monomer:	Vinyl acetate (ethylene or acetylene and acetic acid)
Polymerization:	Free radical initiated chain polymerization
Major Uses:	Adhesives, emulsion paints, paper and textile treatment
Major Producers:	Air Products, Borden, Hoechst, Monsanto, Rhône-Poulenc, Union Carbide

Polyvinyl acetate is the starting material for polyvinyl alcohol, polyvinyl butyral, and polyvinyl formal. It is manufactured primarily by free radical initiated emulsion polymerization. Polyvinyl acetate polymerizes mainly head-to-tail, but some of the monomers orient themselves head-to-head and tail-to-tail as the chain grows. The density of PVA is 1.191 g/cm^3, and the glass transition temperature is 28 to 31 °C. Copolymers with acrylates, maleates, and ethylene are also produced. A random copolymer with ethylene, poly(ethylene-co-vinyl acetate) EVA, has advantages of greater flexibility and greater acceptance of filler over PE.

The vinyl acetate monomer is obtained by reaction of ethylene with acetic acid and oxygen in the presence of palladium chloride:

$$CH_2=CH_2 + CH_3COOH \xrightarrow[PdCl_2]{O_2} CH_2=CH-OCOCH_3$$

An earlier process was based on the reaction of acetylene with acetic acid:

$$CH\equiv CH + CH_3COOH \longrightarrow CH_2=CH-OCOCH_3$$

Polyvinyl acetate is used as an adhesive for packaging and wood gluing. It is also used as a latex for the manufacture of interior and exterior paints. The toughest paint films are formed from latex polymers having the highest molecular weights. Binders for paper coatings and textile finishes are also formulated with PVA.

Polyvinyl acetate films cannot be produced by melt extrusion. Air Products has developed a copolymer (Vinex) consisting of PVAL, PVA, and poly(alkyleneoxy) acrylate units. The long chain reduces the crystallinity of PVA and acts as an internal, non-migrating plasticizer. The thermoplastic copolymer is water soluble, thermally stable, and biodegradable.

Polyvinyl Alcohol (PVAL)

$$\left[CH_2\underset{OH}{CH} \right]_n$$

Monomer: Vinyl acetate
Manufacture: Hydrolysis of PVA
Major Uses: Textile and paper treatment
Major Producers: Air Products (Vinol), Du Pont (Elvanol), Hoechst, Monsanto (Gelvatol), Rhône-Poulenc

Polyvinyl alcohol is manufactured by controlled hydrolysis of polyvinyl acetate using base (sodium hydroxide) catalyzed methanolysis. The physical properties of PVAL are controlled by molecular weight and degree of hydrolysis (usually 88 to 98 %). The specific gravity of PVAL is 1.27 to 1.31 g/cm^3, and the glass transition temperature is 75 to 85 °C. PVAL is water soluble, and it adheres well to cellulose surfaces.

Major uses are in textile sizing, adhesives, paper coatings, joint cements, water-soluble films, and nonwoven fabric binders.

Polyvinyl Butyral (PVB) and Polyvinyl Formal (PVF)

$$\left[\!-CH_2-CH\underset{O}{\overset{CH_2}{\diagdown}}CH-\!\right]_n \quad \left[\!-CH_2-CH\underset{O}{\overset{CH_2}{\diagdown}}CH-\!\right]_n$$
$$\underset{|}{CH}CH_2$$
$$(CH_2)_2CH_3$$

Polyvinyl butyral is produced from butyraldehyde and fully hydrolyzed polyvinyl alcohol in the presence of strong mineral acids. Polyvinyl formal is obtained in a similar manner from polyvinyl alcohol and formaldehyde. The conditions of acetal formation are closely controlled to form a polyvinyl acetal containing predetermined proportions of acetate groups, hydroxy groups, and acetal groups. The hydroxy groups can be used for further crosslinking. PVB is marketed by Du Pont (Butacide) and Monsanto (Butvar). Other producers include Henkel, Hoechst, and Rhône-Poulenc.

By far the largest single use of PVB is as an adhesive in the manufacture of laminated safety glass. PVB dispersions are used in the textile industry to impart abrasion resistance, durability, strength, and slippage control. Polyvinyl formal is used in the manufacture of enamels for heat-resistant wire insulation. Polyvinyl acetals are also used in high performance thermosetting adhesives and in hot-melt formulations.

Fibers based on formaldehyde-modified polyvinyl alcohol are generally known as vinal or vinylon fibers. These products are presently produced in Japan. Vinylon fiber is solution spun using an aqueous PVAL solution, which is coagulated in a concentrated sodium sulfate bath. Heat treatment induces insolubility, and the heat-treated fiber is passed through a formaldehyde bath for further crosslinking. Vinylon fibers have a cottonlike feeling but soil easily. Their major uses are in industrial spun yarns, canvas applications, industrial coverings, and awnings.

Polyvinyl Ethers

$$\left[\!-CH_2-\underset{\underset{OR}{|}}{CH}-\!\right]_n$$

Commercial uses have developed for several alkyl polyvinyl ethers, such as methyl, ethyl, and isobutyl derivatives. The polymerization is usually conducted by cationic initiation, using Friedel-Crafts type catalysts (e.g., boron trifluoride, aluminum trichloride, stannic chloride). The monomers are produced from acetylene and the corresponding alcohols. The major producer of polyvinyl ethers is GAF. Copolymers of methyl vinyl ether and maleic anhydride are marketed under the trade name Gantrez. They are used for latex paints, in cosmetics, and in hair sprays. Copolymers of vinyl ethers with vinyl acetate and vinyl chloride are also known.

The glass transition temperatures and melting points of some polyvinyl ethers are listed in Table 30.

Table 30 Glass Transition Temperature and Melting Points of Polyvinyl Ethers

R	Tg (°C)	M.p. (°C)
Methyl	−34	144
Ethyl	−42	
Isopropyl	−3	191
n-Butyl	−55	
Isobutyl	−19	170
t-Butyl		238
2-Ethylhexyl	−66	
n-Hexyl	−77	
n-Octyl	−80	

The water-soluble polyvinyl ethers have found utility in adhesives, coatings, lubricants, and greases.

Polyvinylpyrrolidone (PVP)

Polyvinylpyrrolidone is produced by free radical initiated bulk, solution, or suspension polymerization of n-vinylpyrrolidone. The monomer can also be polymerized with cationic (BF_3) or anionic (KNH_2) initiators. n-Vinylpyrrolidone is manufactured from acetylene and pyrrolidone. The homopolymer is water soluble. Copolymers with vinyl acetate and other comonomers are also produced.

Polyvinylpyrrolidone is used in cosmetics (hair and skin care), in textile treatment, in adhesives, and as a plasma volume extender. Incorporation of PVP in hydrophobic fibers, such as polyacrylonitrile, polyesters, nylon, and cellulosic fibers increases their dyeability.

Polyvinylcarbazole

Polyvinylcarbazole is also manufactured by free radical or Ziegler-Natta initiated chain polymerization. The monomer is available from acetylene and carbazole.

Polyvinylcarbazole is used in the formulation of photopolymer systems, and it has found utility as an organic photoconductive material in xerography.

12 Polyurethanes

Introduction

The polyolefins are by far the most important addition polymers. In the process of generating the polymer backbone, carbon-carbon bonds are formed, and the process is for all practical purposes irreversible. In contrast, polyurethanes and polyacetal are addition polymers in which reversible adducts are formed. In the case of the polyurethanes, reversal of the generated carbon-oxygen bonds occurs above 120 °C. The reverse process is usually catalyzed by the same catalysts that are used to facilitate polymer formation. Polyacetal, an important engineering thermoplastic, is also thermally labile, and unzipping is observed upon heating. The latter polymer has been stabilized by end-group capping (see Chapter 13). The three groups of addition polymers (see Part I, Chapter 2) are depicted in Figure 33.

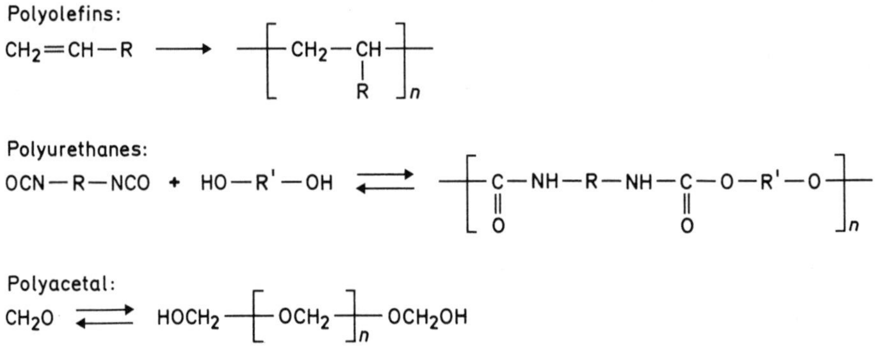

Polyolefins:

$CH_2=CH-R \longrightarrow +CH_2-CH(R)+_n$

Polyurethanes:

$OCN-R-NCO + HO-R'-OH \rightleftharpoons +C(=O)-NH-R-NH-C(=O)-O-R'-O+_n$

Polyacetal:

$CH_2O \rightleftharpoons HOCH_2+OCH_2+_n OCH_2OH$

Figure 33 Industrial addition polymers

Polyurethanes are by far the most versatile group of polymers, because products ranging from soft linear thermoplastic elastomers to hard thermoset rigid foams are readily produced from liquid monomers. Polyurethanes are subdivided into four broad groups: flexible foam, rigid foam, elastomers and adhesives, coatings and sealants. The addition polymerization of diisocyanates with macroglycols to give polyurethanes was

12 Polyurethanes

pioneered by O. Bayer in 1937 at the I. G. Farbenindustrie laboratories in Leverkusen, Germany. The development of polyurethane technology was delayed by World War II. After the war, urethane polymers became known in the United States, where commercial production of flexible foam began in 1953. The availability of low cost polyether polyols in 1957 accelerated the growth of polyurethanes, and polymeric isocyanates were introduced in the late 1950s. By 1988 the global consumption of polyurethanes amounted to 4,666,000 metric tons.

The basic building blocks for polyurethanes are di- and polymeric isocyanates and macroglycols, also called polyols (see Chapter 13). The commonly used isocyanates are tolylene diisocyanate (TDI) and methylenediphenyl diisocyanate (MDI) and polymeric isocyanate (PMDI) mixtures manufactured by phosgenating polyamines obtained in the acid-catalyzed condensation of aniline and formaldehyde. MDI and PMDI are coproducts, and separation is achieved by distilling part of the MDI from the reaction mixture. The reactions used in the manufacture of commercial isocynates are summarized in Figure 34.

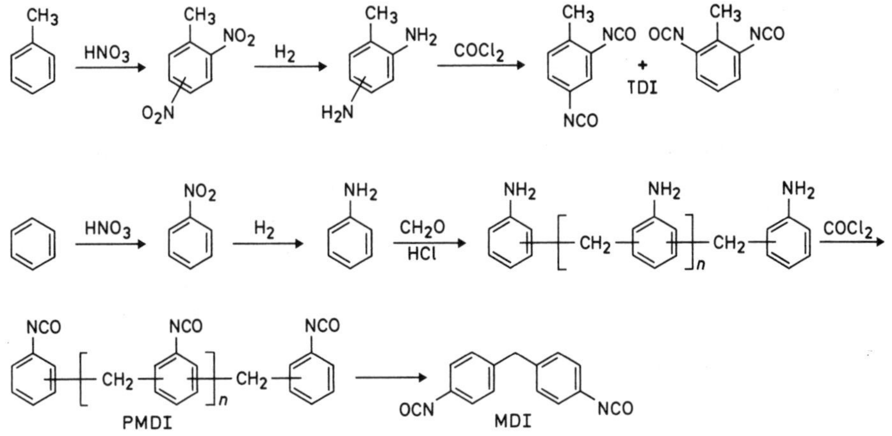

Figure 34 Manufacture of commercial isocyanates

Specialty aliphatic isocyanates, such as hexamethylene diisocyanate, isophorone diisocyanate, and H_{12}MDI are used in light-stable coatings.

The macroglycols used as comonomers are either polyether or polyester based. Polyether diols are obtained by ring-opening polymerization of alkylene oxides (see Chapter 13), and commonly used polyester diols are polyadipates. A polyol produced by ring-opening polymerization of caprolactone initiated with low molecular weight glycols, is also used. The reactions used in the formation of the macroglycols are summarized in Figure 35.

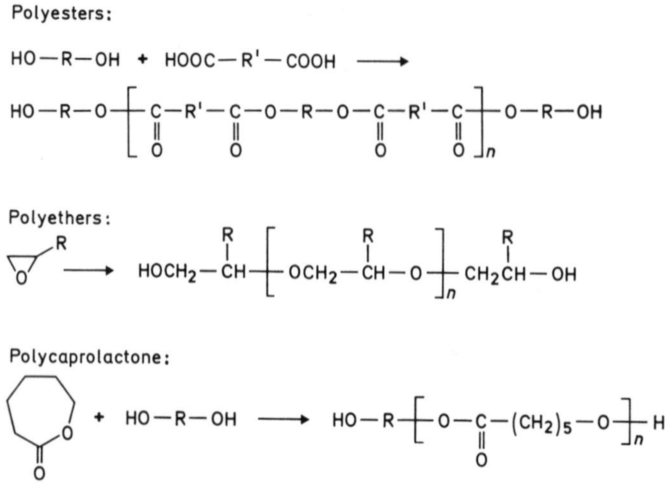

Polyesters:

$$HO-R-OH + HOOC-R'-COOH \longrightarrow$$

$$HO-R-O\left[\begin{array}{c}C-R'-C-O-R-O-C-R'-C\\ \| \quad \| \quad \quad \| \quad \| \\ O \quad O \quad \quad O \quad O\end{array}\right]_n O-R-OH$$

Polyethers:

$$\overset{R}{\underset{O}{\triangle}} \longrightarrow HOCH_2-\overset{R}{\underset{|}{CH}}\left[\begin{array}{c}OCH_2-\overset{R}{\underset{|}{CH}}-O\end{array}\right]_n CH_2\overset{R}{\underset{|}{CH}}-OH$$

Polycaprolactone:

$$\text{(lactone ring)} + HO-R-OH \longrightarrow HO-R\left[\begin{array}{c}O-C-(CH_2)_5-O\\ \| \\ O\end{array}\right]_n H$$

Figure 35 Manufacture of macroglycols

Flexible polyurethane foams are manufactured from TDI and higher molecular weight polyether triols. Water is used as the blowing agent in the manufacture of flexible foam. The polymers obtained have a polyurethane urea structure that is due to the reaction of the diisocyanate with water. Low density flexible foams are used in furniture and bedding applications, higher density flexible foams are used in automotive seating applications, and semiflexible foams are used for automotive interior padding.

Low density rigid foams are manufactured from PMDI and higher functional polyether or aromatic polyester polyols. These foams are the most efficient commercially available insulation materials, and they are widely used in the construction and transportation industries. High density rigid polyurethane foams are used in structural parts. The chlorofluorocarbons (CFCs) used as blowing agents in the manufacture of rigid polyurethane foam insulation are being phased out by the end of the century, which mandates the rapid development of commercially viable blowing agent that do not contribute to ozone depletion.

Polyurethane elastomers have the rigidity of plastics and the resiliency of rubber, and they are often referred to as elastoplastics. Polyurethane elastomers are used in the automotive industry for the manufacture of exterior parts by reaction injection molding (RIM). The reaction injection molding process has economic advantages over the injection molding of engineering thermoplastics and their alloys. Elastomer-type formulations are also used in spandex fibers and in adhesives, coatings, and sealants.

Flexible Polyurethane Foam

Monomers:	TDI, polyether polyols, water
Polymerization:	Bulk addition polymerization
Major Uses:	Furniture and bedding (66 %), transportation (29 %), carpet underlay, packaging
Major Producers:	Isocyanates: BASF (TDI), Bayer (Mondur T-80), Dow (Voranate T-80), ICI (Rubinate TDI), Olin (TDI-80)
	Polyols: BASF (Pluracol), Dow (Voranol), Mobay (Multranol), Olin (Poly-G), Union Carbide (Niax)

Most flexible foam is made from TDI (80/20 mixture of isomers indicated in Figure 34) and polyether triols with a molecular weight of about 3000. Trifunctionality is required to produce a network-type, three-dimensional polymer. Flexible foams are water blown; that is, the reaction of water with the isocyanate produces CO_2 gas, which acts as a blowing agent, producing the open-cell foam structure. Sometimes fluorocarbon or methylene chloride is added for greater softness and lower density. Furniture grade flexible polyurethane foams have a density of 0.024 g/cm^3. Higher density (0.045 g/cm^3) foams, so-called high resiliency foams, use polyether triols with a molecular weight of 6000 or polymer polyols (see Chapter 13) to improve the load-bearing properties.

Flexible slab or bun foam is typically poured by multicomponent machines at rates exceeding 45 kg/min. So-called one shot pouring from traversing mixing heads is generally used. The scrap generated in the manufacture of flexible foam buns is used as carpet underlay. The flexible foam used in the transportation industry is made by molding. In this manner seat cushions, back cushions, and bucket-seat padding are obtained. For flexible foam molding, TDI, PMDI, or blends of TDI/PMDI are often used in conjunction with reactive polyether triols with molecular weights of 4500 to 6500 having a primary hydroxyl content of greater than 50 %.

The largest volume use of flexible foam is in furniture and bedding. Almost all furniture cushioning is flexible polyurethane foam. The second largest market for flexible foam is transportation (i.e., seats for passenger cars, other motor vehicles, and airplanes). Flexible and semirigid polyurethane foam products are also used in packaging.

Rigid Polyurethane Foam

Monomers:	Polymeric MDI (PMDI), polyether, and polyester polyols
Polymerization:	Bulk addition polymerization
Major Uses:	Building and construction (68 %), refrigeration (18 %), transportation, packaging
Major Producers:	Isocyanates: BASF (Lupranate M-20S), Dow (Papi), ICI (Rubinate M), Mobay (Mondur MR)
	Polyols: Similar to flexible foam

Rigid polyurethane foam is mainly used for insulation. The configuration of the final product determines the method of production. Rigid polyurethane foam is produced in slab or bun form on continuous lines similar to flexible foam, or it is continuously laminated between asphalt or tar paper, aluminum, steel, fiberboard, or gypsum facings. It can be poured or frothed into suitable cavities or sprayed onto suitable surfaces.

Almost all rigid polyurethane foam is produced from PMDI. Polyols of choice include propylene adducts of higher functional initiators (amino alcohols, diamines, oligoamines, pentaerythritol, sorbitol, sucrose, phenol-formaldehyde resins), and aromatic polyester diols. The amine-initiated polyols have autocatalytic activity because of their tertiary amino groups, and they are preferred in molding and spray applications. The crude aromatic polyester diols are used in combination with the multifunctional polyols. The high functionality of the polyether polyols combined with the higher functionality of PMDI (2.7) contributes to the rapid network formation required for rigid polyurethane foams. Fire retardants containing phosphorus or halogen are sometimes added.

Insulation foams are chlorofluorocarbon blown. At a density of 0.032 g/cm^3, rigid polyurethane foam is about 97 vol% gas (CFC 11), which accounts for the observed low K-factor. CFC 11(trichlorofluoromethane) has the lowest thermal conductivity of all known gases suitable for foam expansion and suffers little outward diffusion from the foam cells. Currently, attempts are under way to replace CFC 11 with other blowing agents: the manufacture of CFC 11 will be phased out by the end of the century because of ozone depletion allegedly caused by the CFCs. Hydrogen-bearing hydrohalocarbons are considered as replacements, because they are oxidized in the lower atmosphere, where they react before reaching the ozone layer. However, the leading candidate, hydrochlorofluorocarbon-123 (CF_3CHCl_2), was found to cause benign tumors in male rats, which may prevent its use. Another possible blowing agent could be dimethyl ether, which can be readily manufactured from synthesis gas.

Urethane-modified isocyanurate foams are also used in insulation applications. These foams are made by using an excess of PMDI, some polyol, a blowing agent, and a suitable isocyanate trimerization catalyst. Urethane-modified isocyanurate foams exhibit superior thermal stability and combustibility characteristics.

The bulk of the rigid polyurethane and polyisocyanurate foams are used in building and construction insulation, and in the insulation of refrigerators, coolers, pipes, truck trailers, railroad freight cars, and cargo containers. Spray applications include building, roof, and tank insulation. Other applications for rigid polyurethane foams are packaging and molded furniture (simulated wood ceiling beams).

Polyurethane Elastomers

Monomers:	MDI, TDI, butanediol, polyether, and polyester diols
Polymerization:	Bulk addition polymerization
Major Use:	Transportation, footwear, sporting goods, spandex fiber
Major Producers:	Dow (Rimthane, Pellethane), Du Pont (Lycra), B. F. Goodrich (Estan), Mobay (Desmopan, Texin)

The polyurethane elastomers are subdivided into thermoplastic (linear) and thermoset (crosslinked) materials. Initial development of polyurethane elastomers concentrated on products similar in processing to natural rubber. These so-called millable polyurethanes are compounded on a rubber mill or by heavy-duty mixers and are generally cured by vulcanization. Subsequently liquid thermoset cast systems with high tensile and tear strength were developed, followed by the introduction of linear segmented thermoplastic polyurethane elastomers (TPUs), which can be processed by injection molding or extrusion. More recently, high pressure impingement mixing machines became available, which led to the production of cast thermoset elastomers by reaction injection molding (RIM).

Millable polyurethane elastomers are produced by chain extension of linear or lightly branched polyester or polyether polyols with aromatic diisocyanates, such as TDI and MDI. Millable elastomers are generally cured by vulcanization using either sulfur- or peroxide-initiated crosslinking. Cast elastomers are crosslinked thermoset polymers produced by liquid casting. They are based on TDI (Adiprene L) or on MDI. In reaction injection molding, liquid forms of MDI or liquid MDI prepolymers are used in combination with the polyols and glycol extenders. To obtain faster reaction rates and better green strength, an aromatic diamine extender such as diethyl TDA was used instead of the glycol extenders. This new family of RIM products is named polyurethane urea, and most systems used in the automotive industry today are based on this chemistry. In 1984, RIM polyurethane systems with an internal mold release (IMR) were introduced, greatly reducing the need to apply external mold release agents. To obtain even faster reaction rates and improved thermal stability, polyurea RIM systems were developed. In polyurea RIM systems the polyols were replaced by amine-terminated polyols (polyether amines), and also aromatic diamine extenders are used in conjunction with MDI. The short reaction times (0.8 to 1.2 seconds) limit polyurea RIM systems to small parts or to special equipment with high injection rates.

The RIM elastomers are used in automotve fascia, bumper, and fender extensions. More recently RRIM (glass-reinforced RIM) has been used for fenders, door panels, and rear quarter panels. Low modulus systems are used for glass encapsulating gaskets for trunk and car windows and windshields.

TPUs are segmented linear thermoplastic polymers based on MDI, polyester, or polyether diols and glycol extenders. The polyol constitutes the soft segment, whereas the hard segments are formed from the reaction of MDI with the extender glycol. These segmented block copolymers are sometimes called domain polymers, because the hard blocks aggregate in domains held together by hydrogen bonding. Commonly used soft segments include polyester diols, such as adipates, and polycaprolactone, as polyether diol polytetramethylene glycol is used. The hard segments are formed from MDI and short-chain glycol extender, such as 1,4-butanediol, 1,6-hexanediol, ethylene glycol, or diethylene glycol. Polyether-based TPUs have superior hydrolytic stability, whereas polyester-based products have better oil resistance. The polymers can be formulated with isocyanate or hydroxyl end groups. For extrusion grade materials, lower molecular weight resins terminated by hydroxyl groups are preferred because of their lower melting range.

Thermoplastic polyurethane elastomers are molded or extruded to produce elastomeric products used in automotive parts, shoe soles, sport boots, roller skate and skateboard wheels, disposable diaper bands, pond liners, tubings, cable jackets, gears, and other mechanical goods.

Du Pont's elastomeric polyurethane spandex-type fiber (Lycra) was introduced in 1962. The generic name "spandex fibers" designates elastomeric fibers in which the fiber-forming substance is a long-chain polymer consisting of more than 85 % segmented polyurethane. The soft block again is a macroglycol, while the hard block is formed from MDI and hydrazine or ethylenediamine.

$$HO\sim\sim\sim OH \quad + \quad OCN-R-NCO \longrightarrow$$

Macroglycol M.W. 2000   MDI

$$OCN-R-NHCOO\sim\sim\sim OCONH-R-NCO \xrightarrow[\text{Chain extender}]{H_2NNH_2}$$

Prepolymer

$$-\!\!\left[COO\sim\sim\sim OCONH-R-NHCO-NHNH-CONH-R-NH\right]_n\!\!-$$

Soft segment   Hard segment

R = –C₆H₄–CH₂–C₆H₄–

Lycra is soluble in *N,N*-dimethylformamide and the fiber is extruded through a spinnerette into a column of circulating hot air (dry spinning). Elastomeric spandex fibers are used in hosiery and sock tops (40 %), in girdles, bras, and support hose (25 %), and in lightweight knitted swimwear. The use of spandex fibers in outerwear clothing, especially slacks, jeans, and sportswear, is increasing.

Polyurethane Coatings, Adhesives, and Sealants

Monomers:	MDI, TDI, butanediol, polyester, or polyether polyols
Polymerization:	Addition polymerization
Major Uses:	Industrial coatings, transportation, construction, footwear
Major Producers:	BASF, Bayer, B.F.-Goodrich, ICI, PPG, Sherwin-Williams

Polyurethane systems are also formulated for coating applications using either finished polymers, prepolymers, or two-component systems. The finished polymers can be applied from polar solvents or by hot-melt or powder coating technologies. Prepolymers can be one- or two-component systems. The one-component system requires blocking of the isocyanate groups to prevent curing in the container. Examples of blocking agents are given in Table 31.

Table 31 Blocked Isocyantes

Blocking Agent	Unblocking Temperature (°C)
Phenol	160
m-Nitrophenol	130
Diethyl malonate	130–140
Acetone oxime	180
Caprolactam	160
Hydrogen cyanide	120–130

The disadvantage of blocked one-component systems is the generation of the blocking agent upon curing. A masked aliphatic diisocyanate system is shown in Figure 36. In this case the cyclic bisurea derivative is stable in the polyol or in a water emulsion formulated with the polyol. Upon heating, ring-opening occurs, generating the diisocyanate, which instantaneously reacts with the macroglycol to form a polyurethane. Aliphatic diisocyanates based polyurethanes are light stable. In contrast, coatings made from TDI or MDI gradually discolor upon exposure to light and oxygen.

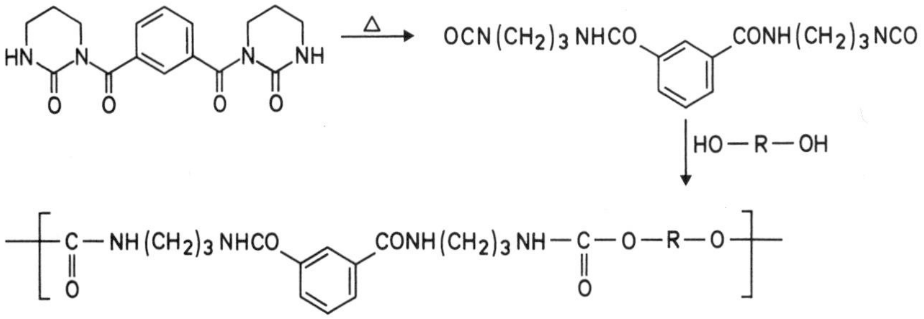

Figure 36 Macrocyclic ureas as masked diisocyanates

Polyurethane power coatings are formulated from polyester polyols and blocked polyisocyanate crosslinking agents. Most commercial crosslinking agents are based on caprolactam-blocked TDI and IPDI. Powder coatings contain pigments, fillers, and small amounts of flow-control additives. The powder coatings are cured at 180 to 200 °C for 15 to 30 minutes using stannous octoate as the catalyst.

Moisture-cured polyurethane coatings are isocyanate-terminated prepolymers which, after application, are cured by the reaction of the residual isocyanate groups with moisture. The amino groups initially formed react with more isocyanate to form urea linkages. The isocyanates used in the two-component coatings include TDI-trimethylolpropane adducts, HDI-biurets, and HDI-isocyanurates. The derived systems cure at ambient temperature to give coatings with excellent physical properties. Such coatings are applied as industrial finishes.

Water-based polyurethane coatings are being developed to reduce solvent emission. Some suppliers provide aqueous polyurethane dispersions intended for coatings on plastics, metal, wood, concrete, rubber, paper, and textiles. Self-crosslinking aqueous polyurethane dispersions have been developed recently. These coatings are intended for aircraft, appliances, automotive components, and industrial and farm machinery. Polyurethanes must be modified to form stable aqueous dispersions. Examples include ionic polymers, where the ionic group acts as internal emulsifier, and nonionic polymers with attached hydrophilic polyether chain segments.

Polyurethanes are also widely used in adhesive applications. Polyisocyantes, modified polyisocyanates, prepolymers with terminal isocyanate groups, soluble polyurethane elastomers, and aqueous dispersions are used in a wide variety of adhesive applications. Again, one- or two-component systems are used. The one-component reactive adhesives are isocyanate-terminated prepolymers that cure by reaction with moisture. A typical one-component elastomeric adhesive resin consists of a linear polyester (polybutylene adipate), MDI, and 1,4-butanediol. Crystallization of the soft polyol segment below 40 °C would impair the physical properties of most elastomers in the main temperature range of use. Crystallization is beneficial because it contributes to bond strength. One-component elastomeric adhesives are used for the manufacture of laminated films for packaging. The two-component reactive adhesives are formulated from polyisocyanates and polyols. They are used as structural adhesives, especially for the bonding of plastic automotive parts.

Another important application of polyurethanes consists of architectural sealants. In this application all the foregoing technologies are utilized. Most sealats are soft resilient elastomers but they can also be modified with epoxy, acrylics, or other polymers. They are available as pourable (self-leveling) and gunnable (nonsag) types. Polymeric MDI is also used as a binder for wood products, such as particle board, wafer board, and chip board, and as a core binder for binding of sand molds used for metal casting. Also athletic surfaces are sometimes prepared from rubber tire scrap bonded with isocyanate prepolymers, and flexible polyurethane scrab is bonded with isocyanate prepolymers to form rebounded foam used as carpet underlay.

13 Ether Polymers

Introduction

Many linear polymers have ether linkages in their backbone structure. They are manufactured by a variety of polymerization processes, such as polyaddition (polyacetal), ring-opening polymerization (polyethylene oxide, polypropylene oxide, and epoxy resins), oxidative coupling (polyphenylene oxide), and polycondensation (polyether sulfone, polyether imide). Polyether sulfones are treated in Chapter 17, and polyether imide is described in Chapter 15. Polycarbonates and polyarylates with ether backbones are discussed in Chapter 14.

Polyacetal and polyphenylene oxide (PPO) are widely used as engineering thermoplastics; thermoset epoxy resins are used as coatings and adhesives; and polyethylene oxide and polypropylene oxide are used mainly as macroglycols in the production of polyurethanes. Often, block copolymers are used for this purpose because primary hydroxyl groups, obtained from an ethylene oxide end block, are more reactive with aromatic diisocyanates than polypropylene oxide derived products, which contain primary and secondary hydroxyl end groups.

Polyacetal $-\!\!+\!\!OCH_2\!\!+\!\!_n\!\!-$

Monomer:	Fomaldehyde, ethylene oxide
Polymerization:	Cationic or anionic chain polymerization
Major Uses:	Appliances, plumbing and hardware, transportation
Major Producers:	BASF (Ultraform), Du Pont (Delrin), Hoechst Celanese (Hostaform, Celcon)

Polyoxymethylene (polyacetal) is the acetal of formaldehyde, and it is obtained by polymerization of aqueous formaldehyde or by ring-opening polymerization of trioxane, the cyclic trimer of formaldehyde. The ring-opening reaction is the preferred method. The bulk polymerization of trioxane is conducted with cationic initiators. In contrast, highly purified formaldehyde is polymerized in solution using either cationic or anionic initiators. End-group capping is essential; otherwise depolymerization ("unzipping") would occur under molding conditions. The capping of the end groups can be achieved by etherification or, more conveniently, by esterification using acetic anhydride.

trioxane $\longrightarrow$ $HOCH_2\!\!+\!\!OCH_2\!\!+\!\!_n\!\!OCH_2OH$ $\xrightarrow{Ac_2O}$ $CH_3COOCH_2\!\!+\!\!OCH_2\!\!+\!\!_n\!\!OCH_2OOCCH_3$

If the polymerization is conducted in the presence of ethylene oxide, end-group capping is not necessary. Treatment of the copolymer with aqueous alkali hydrolyzes the product to give stable hydroxyethyl end groups.

Linear polyoxymethylene has been known for over a hundred years, but a stable product was first developed in 1958 by Du Pont. The copolymer with ethylene oxide (Celcon) was commercialized by Celanese in 1961. BASF markets a similar copolymer under the trade name Ultraform.

Polyacetal is obtained as a mainly crystalline linear polymer, with an average molecular weight of 30,000 to 50,000. It is stable in air up to 100 °C, and its high mechanical strength over a wide temperature range, combined with good electrical properties, renders this material useful as an engineering thermoplastic. The unfilled materials are hard, strong, and stiff and have good toughness. The homopolymer exhibits a heat deflection temperature of 124 °C at 264 psi, and the maximum recommended continuous use temperature is 85 °C. The copolymer exhibits a heat deflection temperature of 110 °C at 264 psi, and a recommended maximum use temperature of 104 °C. Impact-modified grades are formulated by using polyurethane elastomer blends. Polyacetals are used widely in the molding of telephone components, radios, small appliances, fuel system components, pump housings, garden hose nozzles, and mechanical parts.

Polyacetal is one of the very few plastic materials not based on petrochemical feedstocks. Methanol, the precursor of formaldehyde, can be obtained from natural gas or coal via synthesis gas at competitive costs.

Other engineering plastics include engineering grades of ABS, nylons, polycarbonate, PPO, PPS, PET, PBT, polyarylates, polyketones, and polysulfones (see also Chapter 17). These products are often reinforced or filled to improve stiffness (moduli) and thermal properties and to improve its cost. The primary areas of use for the engineering plastics are transportation, and electrical and electronic products. The largest factor in their remarkable growth over the years has been and will be their spectacular inroads as metal replacements.

Polyethylene Glycol (PEG) and Polypropylene Glycol (PPG)

$$HOCH_2CH_2-[OCH_2CH_2]_n-OH \qquad HOCHCH_2-[OCH_2CH]_n-OH$$
$$||$$
$$CH_3 CH_3$$

                PEG   PPG

Monomer:	Ethylene oxide, propylene oxide
Polymerization:	Base-catalyzed, ring-opening
Major Uses:	Polyols for polyurethanes (> 80 %), surfactants, lubricants
Major Producers:	BASF (Pluracol), Dow (Voranol), ICI (Daltorez), Mobay (Multranol), Olin (Poly-G), Union Carbide (Niax)

13 Ether Polymers

Polyether polyols are addition products of cyclic ethers. The polymerization is initiated by adding the alkylene oxide to the corresponding glycol or water in the presence of potassium hydroxide as catalyst. The materials are nonvolatile and range from viscous liquids to waxy solids depending on type and molecular weight. The polyols used in polyurethanes have an average molecular weight ranging from 200 to 10,000. Block copolymers are also produced. For example, propylene oxide is first reacted with propylene glycol to form the homopolymer. Further reaction with ethylene oxide produces a block copolymer. The advantage of the copolymers is the generation of primary hydroxyl end groups. Polyols with primary hydroxyl end groups are more reactive with isocyanates. In addition to the block copolymers, random copolymers are produced by feeding a mixture of the alkylene oxides. A statistical distribution is not obtained because ethylene oxide is more reactive. If equal amounts of ethylene oxide and propylene oxide are used the chains are terminated primarily in propylene units having secondary hydroxyl end groups.

The ether polyols obtained in this manner are not truly difunctional because of formation of monoalcohols in propylene oxide derived polyols. Therefore thermoplastic polyurethane elastomers cannot be made from propylene oxide derived polyols. However, the major uses of propylene oxide derived polyols are in flexible polyurethane foams, where a functionality of approximately 3 is required. In this case glycerol, a trifunctional alcohol, is used as the initiator. Higher functional polyols used in rigid polyurethane applications are made by using higher functional initiators as indicated in Table 32. The polyols using an amine initiator are more reactive in their reaction with isocyanates because of the catalytic effect of the tertiary amino groups.

Table 32 Commercial Polyether Polyols

Product	Functionality	Initiator	Alkylene oxide
Poly ethylene glycol PEG	2	Water or ethylene glycol	Ethylene oxide
Poly propylene glycol PPG	2	Water or propylene glycol	Propylene oxide
PPG/PEG*	2	Water or propylene glycol	Propylene/ethylene oxide
Glycerol adduct	3	Glycerol	Propylene oxide
Trimethylolpropane adducts	3	Trimethylolpropane	Propylene oxide
Pentaerythritol adducts	4	Pentaerythritol	Propylene oxide
Ethylenediamine adducts	4	Ethylenediamine	Propylene oxide
Diethylenetriamine adducts	5	Diethylenetriamine	Propylene oxide
Sorbitol adducts	6	Sorbitol	Propylene oxide
Sucrose adducts	6	Sucrose	Propylene oxide

* Random or block copolymers.

Specialty polyols, so-called polymer polyols, containing solid, organic fillers in dispersed distribution, are also produced. The dispersed phase is produced by graft polymerization of olefins, such as acrylonitrile-styrene, directly onto the polyether.

These products are primarily used in the production of elastic soft polyurethane foams with high load-bearing strength. In addition to the graft polymer the polyol can contain the homopolymer of the olefin dispersed in the unaltered polyether. A second approach to filled polyols is offered by polyurea polyols. In this case a diamine is reacted with a diisocyanate in the ether polyol to form interreactive dispersions. In part, also reaction with the hydroxyl end groups occurs. Stable interreactive dispersions obtained in this manner are known as PHD polyethers. The concentration of solids in PHD polyols is limited by viscosity, and a solid content of 20 to 40 % is generally obtained.

A specialty polyether polyol is polytetramethylene glycol (PTMG), which is obtained from tetrahydrofuran by ring-opening polymerization using Lewis acid catalysts:

$$\text{THF} \xrightarrow{\text{Lewis acid}} HO(CH_2CH_2CH_2CH_2O)_n H$$

Polytetramethylene glycol has a functionality of 2 and it is used in the manufacture of polyurethane and polyester elastomers (see Chapters 12 and 14).

Most of the polyether polyols are used in the manufacture of polyurethane foams, elastomers, and coatings, but they are also used in nonurethane applications, such as surfactants, lubricants, functional fluids, thickeners, and plasticizers.

Epoxy Resins

Monomers:	Bisphenol A, epichlorohydrin
Polymerization:	Ring-opening polyaddition reaction with polyfunctional curing agents
Major Uses:	Surface coatings (44 %), laminates and composites (18 %), molding (9 %), flooring (6 %), adhesives (5 %)
Major Producers:	BASF (Epoxin), Bayer (Levepox), Ciba Geigy (Araldit), Dow (DER), Hoechst (Hostapox), Reichhold (Epotuf), Shell (Epon, Epikote)

Eoxy resins were first synthesized in 1934 by P. Schlack, the inventor of nylon-6. The important curing reactions with polyfunctional amines and anhydrides were discovered by P. Castan in Switzerland in 1943, and the products were introduced by Ciba toward the end of World War II. Epoxy resins were initially developed as structural adhesives. The need for bonding of metal was established in the aircraft industry in England, and in 1944 the first successful metal-bonding adhesive was applied in the manufacture of De

13 Ether Polymers

Havilland Hornet twin-engine fighter planes. The first structural synthetic adhesive was based on a phenolic resol combined with a powdered polyvinyl formal resin. High curing temperatures were necessary, and evolution of volatile materials also required the use of pressure. In contrast, epoxy resins cure at room temperature, produce no volatile materials, and cause only a minimum of shrinkage.

Commercial epoxy resins are based on bisphenol A, which upon reaction with epichlorohydrin produces bisglycidyl ethers, were n varies from essentially zero (liquid) to about 25 (hard and tough solid). However, other bisglycidyl ethers and esters and bisepoxides, derived from readily available raw materials, are used also. Examples are shown in Figure 37.

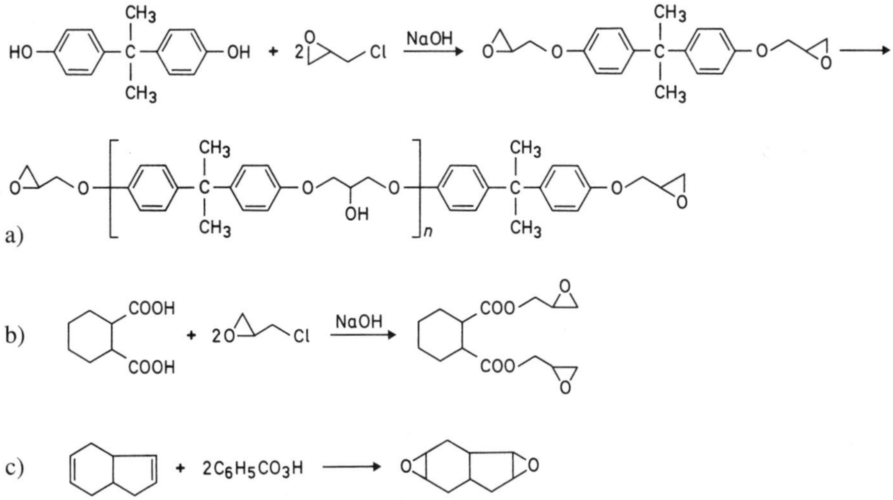

Figure 37 Formation of epoxy resins
(a) Bisphenol A glycidyl ethers, (b) Hexahydrophthalic acid diglycidyl ester
(c) Dicyclopentadiene diepoxide

Expoxies based on aliphatic building blocks have greatly enhanced weatherability. Direct reaction of epichlorohydrin with bisphenol A in high ratios produces liquid resins, while reaction at lower ratios give solids. The molecular distribution of products is easily determined by high-pressuree liquid chromatography. Epoxy resins derived from tetrabromobisphenol A are used in the formulation of flame-retarded systems.

Epichlorohydrin is commercialy produced from propylene and chlorine by the following reaction sequence:

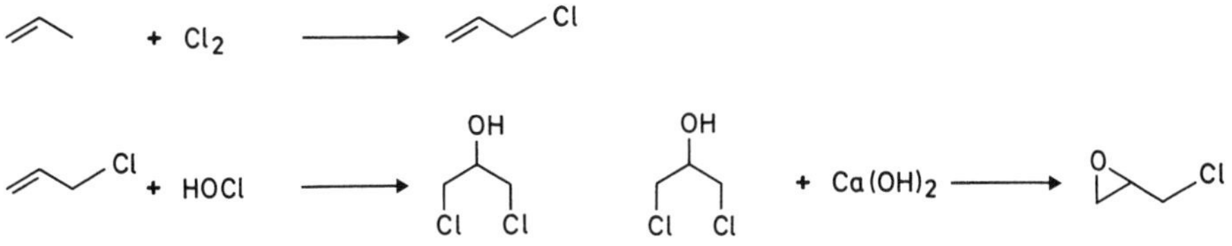

Curing of epoxy resins with di- and polyfunctional active hydrogen-containing molecules, such as polyamines, polyamides, polyacids, polymercaptans, and polyphenols, produces hard and tough thermoset network polymers with excellent adhesion to many substrates. A wide variety of curing agents are used, and liquid amines, such as triethylenediamine, are preferred. For very fast cure at room temperature, accelerators, such as polymercaptans, are added. Sometimes nonreactive diluents, such as dioctyl phthalate, or reactive long-chain monoepoxides, are added to the formulation. High molecular weight compounds can also be added to increase toughness. Nitrile rubber, amorphous nylon, and phenol-formaldehyde and urea-formaldehyde resins are examples.

The ratio of diepoxide to curing agent is important, because crosslinked thermosets are obtained only if stoichiometric amounts of reagents are used. Prepolymers with epoxy end groups are produced if an excess of the diepoxide is used. Table 33 shows the products to be expected from various proportions of diepoxide and tetramine.

Table 33 Effect of Stoichiometry on Product Formation

Bisepoxide	Tetramine	Product
4	1	Epoxy-terminated prepolymer
2	1	Network polymer (thermoset)
1	1	Linear polymer (thermoplastic)
1	2	Amine-terminated prepolymer

Epoxy resins have found a wide range of applications mainly because of the versatility of the reaction polymerization system. Proper selection of resin, modifier, and crosslinking agent allows tailoring of properties for the cured products. This versatility has been a major factor in the steady growth rates of epoxy resins over the years. The main attributes of epoxy resins are as follows:

- Outstanding adhesion to a variety of substrates, especially metals and concrete
- Excellent chemical resistance
- Very low shrinkage on cure
- High tensile, compressive, and flexural strengths
- Excellent electrical insulation properties
- Resistance to corrosion
- Ability to cure over a wide temperature range

In view of these properties, it is not surprising that about half the amount of epoxy resin used today is in protective coatings applications. Two types of coatings are formulated, those cured at ambient temperature and those that are heat cured. The first types are crosslinked through the epoxide ring, using polyamines, polyamides, polymercaptans, and catalytic cures. In contrast, heat-crosslinked systems are cured through reaction of the hydroxyl groups, using anhydrides and polycarboxylic acids as well as formaldehyde resins. Typical coatings applications for phenol-formaldehyde resin modified epoxies include beverage and food can coatings, drum and tank liners, internal coatings for pipes, wire coating, and impregnation varnishes. Urea-formaldehyde resin modified epoxies have better color and are used as can linings, appliance primers, and coatings for hospital and laboratory furniture. Epoxy coatings have another advantage in that they can be formulated as powder coatings, thus completely eliminating the use of solvents.

Another important application for the appliance and automotive industries consists of cathodic electrodeposition coatings. To improve the coalescence properties of the final coatings, chain-extendend epoxy resins are used. For example, lower molecular weight epoxy resins are coupled with diols, such as polypropylene glycol, polycaprolactone diols, and adducts of bisphenol A with ethylene oxide. Blocked isocyanates are often used as crosslinkers (curing agents) in this application.

The end uses for structural or noncoating applications of epoxy resins are in printed circuit boards made from laminates of glass fiber cloth and epoxy resin, in binders for aggregate in the surfacing of floors subjected to heavy wear, in compounds for patching concrete highways, and in potting and encapsulation of electrical equipment.

The most commonly known adhesive applications involve the two-component liquids or pastes, which cure at room or elevated temperatures.

The total global market for epoxy resins in 1988 was 648,000 tons.

Polyphenylene Oxide (PPO)

Monomers:	2,6-Dimethylphenol, 2,3,6-trimethylphenol
Polymerization:	Condensation polymerization by oxidative coupling
Major Uses:	Automotive, appliances, business machine housings, electrical parts
Major Producers:	Borg-Warner (Prevex), G.E. (Noryl)

Polyphenylene oxide was developed in 1956 by General Electric, and production started in 1960. It was soon realized that the homopolymer was too difficult to process, and modified products containing high impact polystyrene or butadiene-styrene copolymers are marketed exclusively today. The modified materials have improved impact resistance. Blends with crystalline nylons improve the solvent resistance of the modified PPO. The PPO provides excellent heat resistance and toughness, while the crystalline phase provides resistance to solvents. These materials were developed for automotive body panels.

PPO is manufactured by oxidation of 2,6-dimethylphenol in solution using copper salts and pyridine as catalysts. This is a polycondensation process because water is generated in the reaction.

The monomer is obtained by the alkylation of phenol with methanol. End-group stabilization with acetic anhydride improves the oxidative resistance of PPO.

Polyphenylene oxide is an amorphous thermoplastic material with a glass transition temperature of about 210 °C. It is stable up to 150 °C, while maintaining its strength properties and fire resistance. However, oxidation of the methyl groups prevents use of the polymer above 125 °C for long periods of time. PPO has outstanding hydrolytic stability and one of the lowest water absorption rates among the engineering thermoplastics.

Automotive uses of modified PPO and its blends include instrument panels, seat backs, rear spoilers, wheel covers, and connectors. Business machine housings, keyboard bases, printer bases, power tool housings, and appliances are other major applications.

Part III
Condensation Polymers

14 Polyesters

Introduction

Polyesters were historically the first synthetic condensation polymers. They were studied by Carothers and his coworkers in the early 1930s. Condensation polymers are macromolecules in which the repeating unit contains fewer atoms than the monomer or comonomers. In other words, in the polycondensation reaction a by-product is generated, which has to be removed as the reaction progresses. An exception is the formation of condensation polymers by ring-opening polymerization. An example of this reaction is the manufacture of nylon-6 by the ring-opening polymerization of caprolactam (see Chapter 15).

Polyesters are synthesized by a variety of methods, which are summarized in Figure 38. Melt polymerization, ester interchange, and interfacial polymerization (Schotten-Baumann reaction) are used in the manufacture of commercial polyesters.

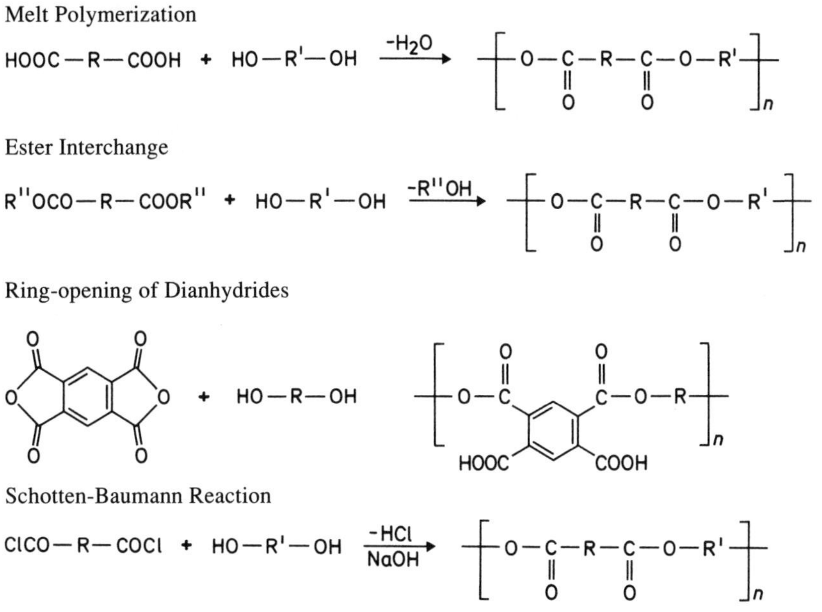

Melt Polymerization

$$HOOC-R-COOH + HO-R'-OH \xrightarrow{-H_2O} {\left[O-\underset{\underset{O}{\|}}{C}-R-\underset{\underset{O}{\|}}{C}-O-R' \right]}_n$$

Ester Interchange

$$R''OCO-R-COOR'' + HO-R'-OH \xrightarrow{-R''OH} {\left[O-\underset{\underset{O}{\|}}{C}-R-\underset{\underset{O}{\|}}{C}-O-R' \right]}_n$$

Ring-opening of Dianhydrides

Schotten-Baumann Reaction

$$ClCO-R-COCl + HO-R'-OH \xrightarrow[NaOH]{-HCl} {\left[O-\underset{\underset{O}{\|}}{C}-R-\underset{\underset{O}{\|}}{C}-O-R' \right]}_n$$

Figure 38   Synthesis of polyesters

The thermoplastic polyesters reached their commercial potential when an economical process for separating the xylene isomers by crystallization was discoverd. The availability of inexpensive *p*-xylene, which is readily oxidized by air to give terephthalic acid, prompted the development of polyester fibers, films, and molding compounds. Du Pont produced polyethylene terephthalate fibers by melt spinning in 1953. Polybutylene terephthalate was commercialized in the early 1970s. The world consumption of major synthetic fibers in 1990 is summarized in Table 34. The total consumption of noncellulosic synthetic fibers was 15,428,000 tons.

Table 34 World Consumption of Synthetic Fibers in 1990

Synthetics	Amount (1.000 tons)
Polyesters	10,184
Polyamides	4,715
Acrylics	2476

Synthetic fibers surpass natural fibers such as cotton and wool in heat stability, durability, resistance to wrinkling, and wash-and-wear properties. The properties necessary to produce good fibers can be summarized as follows: The polymers should be linear, should have a molecular weight of over 10,000 and a high degree of symmetry, should be orientable by drawing, and should contain regularly spaced polar groups to give intramolecular cohesion; in addition, they should have a high melting point and be resistant to heat, water, and chemicals. Preferably, synthetic fibers should also accept dyes.

All major groups of synthetic fibers have certain outstanding features that justify their existence. The major advantages of the polyesters and polyamides are their wash-and-wear properties and their durability. Polyacrylonitrile fibers have self-crimping properties, and rayon fibers have outstanding water absorption characteristics similar to cotton.

Synthetic fibers are spun commercially by melt spinning, which is by far the most economic process, or by solvent spinning. In the latter process, the solvent is removed from the fiber by hot air (dry spinning), or the fiber is precipitated in a nonsolvent (wet spinning). Melt spinning allows the creation of fibers with many different cross sections, and of bicomponent fibers (see Figure 39, below) for specialty applications, such as hosiery.

Expanding on Carother's pioneering work on polyesters, Whinfield and Dixon in England developed polyethylene terephthalate fibers (Dacron, Terylene). The first Dacron polyester plant went into operation in 1953. Polyester fibers are designed to simulate wool, cotton, or rayon, depending on the processing conditions. They have good wash-and-wear properties and resistance to wrinkling, and they recover well from wrinkling. For the first time garments could be worn all day in dry or wet weather and remain relatively wrinkle free. Wrinkles introduced during washing can be removed by tumbling the garment in a dryer. Of course, these properties could be achieved with blends of polyester with cotton and wool also. Blending polyester with resin-treated cotton produces permanent-press fabrics.

Chemical and physical modifications of synthetic fibers further improve bulk and texture. Self-crimping yarns are obtained by combining polymers with different shrinkage characteristics in bicomponent fibers or in mixed shrinkage yarns. Fiber crimp is important in the spinning of staple yarns and in the mechanical behavior of the derived fabric. Cross section modifications provide fabrics with an attractive luster and feel. Some of the cross section options are shown in Figure 39.

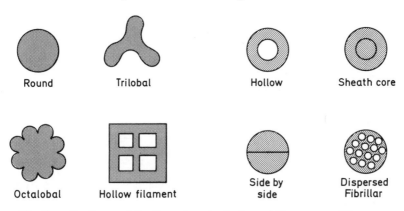

Figure 39 Cross Sections and Configurations in single- and bicomponent fibers

Other applications of multipolymer fibers include antistatic fibers, nylon tire cord with reduced "flat spotting", flame retardance, and improved dyeability; in addition, there are bondable fibers for nonwoven fabrics. Bicomponent fibers with low interfacial adhesion can be fibrillated (split apart) to yield blends of ultrafine fibrils.

Chemical modifications include incorporation of acidic and basic dye receptors and antistatic modifiers.

All commercial polyesters are based on terephthalic acid as the major building block. Variations of the difunctional alcohols as well as the use of alcohol mixtures allows the construction of many different products. However, three major products dominate the market today, polyethylene terephthalate (PET), polybutylene terephthalate (PBT), and polydihydroxymethylcyclohexyl terephthalate (Kodel) (see Table 35).

The difunctional alcohol used in the Kodel-type fibers is also obtained from terephthalic acid. Stepwise reduction of terephthalic acid leads to 1,4-bishydroxymethylcyclohexane. Recently, polyethylene naphthalate was introduced by ICI and Teijin.

Major end uses for polyester fibers in the United States are apparel (61 %), home furnishings (18 %), and tire cord (10 %); uses of the molding grades include bottles (70 %), films, and engineering plastics.

Table 35 Commercial Polyester Fibers

Name	Structure	Density	T_m (°C)	T_g (°C)	Uses
PET	−[−CO−⟨⟩−COO∼O−]$_n$−	1.36–1.38	265	70–80	Apparel, home furnishings
PBT	−[−CO−⟨⟩−COO∿∿O−]$_n$−	1.31	224	40	Home furnishings
Kodel type	−[−CO−⟨⟩−COOCH$_2$−⟨⟩−CH$_2$O−]$_n$−	1.22–1.23	290–195	60–90	Home furnishings

Polyethylene terephthalate (PET)   −[−CO−⟨⟩−COO∼O−]$_n$−

Monomers:	Dimethyl terephthalate or terephthalic acid, ethylene glycol
Polymerization:	Bulk polycondensation
Major Uses:	Fibers, film, bottles, molding compounds
Major Producers:	Fibers: Du Pont (Dacron), Eastman, Fiber Ind. (Angelette), Hoechst (Trevira), Monsanto (Blue C)
	Molding compounds: Akzo (Arnite A), Allied (Petra), Ciba (Crastin), Du Pont (Rynite), G.E. (Valox), Hoechst Celanese (Impet), Mobay (Petlon), Rhône-Poulenc (Rhodester)

Polyethylene terephthalate is widely used in synthetic fibers and in films. Du Pont has introduced a highly crystalline PET with 30 % glass reinforcement (Rynite) as a molding compound. This material has a high heat deflection temperature (227 °C at 264 psi) and it melts at 271 °C (regular PET melts at 265 °C; the melting point is affected by the amount of diethylene glycon by-product present). The glass transition temperature of PET is about 74 °C. Crystallization of PET can be generally achieved upon heating to 190 °C and orientation. In thin film applications, transparency is achieved by rapid quenching with a water-cooled roll.

The production of polyethylene terephthalate is conducted in two steps. Dimethyl terephthalate is heated with ethylene glycol to give a mixture consisting of dihydroxyethyl terephthalate and higher oligomers. Further heating to 270 °C under vacuum in the presence of a catalyst produces the final polymer.

$$CH_3OCO-\underset{}{\bigcirc}-COOCH_3 + HOCH_2CH_2OH \longrightarrow$$

$$HO\sim O-\underset{O}{\overset{\|}{C}}-\underset{}{\bigcirc}-\underset{O}{\overset{\|}{C}}-O\sim OH + \text{Oligomers} \xrightarrow{Sb_2O_3} \text{Polymer}$$

Terephthalic acid is produced by air oxidation of *p*-xylene, and ethylene glycol is obtained from ethylene oxide and water.

The strength of PET in its oriented form is outstanding. Biaxially oriented PET film is used in magnetic tape, in X-ray and other photographic film applications, in electrical insulation, and in food packaging, including boil-in-bag food pouches. Production of PET bottles for carbonated beverages by blow molding has gained prominence because PET has low permeability to carbon dioxide, and it can be easily recycled. Other containers include bottles for toiletries, cosmetics, and household products. Because of its excellent thermal stability PET is also used as coating material for microwave and conventional ovens.

Polybutylene terephthalate (PBT)

$$-\!\!\left[-CO-\underset{}{\bigcirc}-COO\sim\!\!\sim O-\right]_n\!\!-$$

Monomers:	Dimethyl terephthalate or terephthalic acid, butanediol
Polymerization:	Bulk polycondensation
Major Uses:	Automotive exterior parts, electrical applications, small appliances
Major Producers:	Akzo (Arnite T), Atochem (Orgator), BASF (Ultradur), Eastman Kodak (Tenite PTMT), GAF (Gafite), G. E. (Valox), Hoechst (Celanex, Hostadur), Mobay (Pocan), Montedison (Pibiter), Solvay (Arylef)

Polybutylene terephthalate is a semicrystalline engineering thermoplastic. It is produced by transesterification of dimethyl terephthalate with 1,4-butanediol in the presence of tetrabutyl titanate. It is highly crystalline, and its rapid rate of crystallization gives short injection molding cycles. PBT has high mechanical strength, a high heat deflection temperature, low moisture absorption, good dimensional stability, and excellent electrical properties. The continuous use temperature of PBT ranges from 120 to 140 °C. Like all polyesters, PBT has to be dried to a moisture level of less than 0.04 % before processing in order to prevent hydrolytical degradation.

Two types of PBT are produced: one with number-average molecular weights of 23,000 to 30,000, and another with higher molecular weights (36,000 to 50,000). Both products are crystalline, melt at 224 °C, and are generally processed around 250 °C. PBT is also amenable to reinforcement with glass. It processes well in injection molding, blow molding, or extrusion operations.

Major uses include automotive exterior parts, under-the-hood parts, electrical parts, small appliances, lawn mower housings, pump housings, and power tool housings.

Polydihydroxymethylcyclohexyl terephthalate

$$-[CO-\langle\bigcirc\rangle-COO-\frown-\langle\bigcirc\rangle-O]_n-$$

A more hydrophobic polyester fiber was introduced by Eastman Kodak as Kodel in 1958. The sole raw material for this polyester is again dimethyl terephthalate. Reduction leads to a dialcohol, which is used with dimethyl terephthalate in the polycondensation reaction:

$$CH_3OCO-\langle\bigcirc\rangle-COOCH_3 \xrightarrow{H_2} HOCH_2-\langle\bigcirc\rangle-CH_2OH$$

$$HOCH_2-\langle\bigcirc\rangle-CH_2OH + CH_3OCO-\langle\bigcirc\rangle-COOCH_3 \xrightarrow{-CH_3OH}$$

$$-[OCH_2-\langle\bigcirc\rangle-CH_2OCO-\langle\bigcirc\rangle-CO]_n-$$

Polyesters made from cyclohexanedimethanol and cyclohexanedicarboxylic acid are marketed by Eastman under the trade name Ecdel. Also, copolyesters based on cyclohexanedimethanol and mixtures of terephthalic and other dicarboxylic acids are on the market. The primary use for these products is extrusion into film and sheeting for packaging. These amorphous films are brilliantly clear, and they can be used to package hardware and other heavy items.

Ethylene glycol modified copolyesters of the Kodel type are used in blow molding applications to produce bottles for packaging shampoos, liquid detergents, and similar products.

The principle of formation of segmented block copolymers (see Chapters 8 and 12) has also been applied to polyesters, with the "hard" block formed from butanediol and terephthalic acid and the "soft" block obtained from polytetramethylene glycol with a molecular weight of 2000. Variation of the hard block influences the hardness, while the soft segment contributes to the elastomeric porperties. Du Pont introduced these linear polyester elastomers under the trade name Hytrel in 1971.

Cellulose Esters

Monomers: Cellulose, acetic acid, propionic acid, butyric acid
Polymerization: Chemical modification of cellulose
Major Uses: Glazing, tapes, packaging
Major Producers: Bayer (Cellidor), Eastman Kodak (Tenite)

Cellulose esters have been known since 1845, when C. F. Schonbein treated cotton with a mixture of nitric and sulfuric acid. In 1868 J. W. Hyatt used nitrated cellulose with

14 Polyesters

camphor added as platicizer to formulate the first man-made thermoplastic material, which he named celluloid. Celluloid is not used extensively today because of its high flammability.

Thermoplastic materials obtained from cellulose include cellulose nitrate, cellulose triacetate, cellulose acetate, cellulose propionate, cellulose butyrate, and ethyl cellulose. Cellulose acetate is produced by treating cellulose, obtained from wood, with acetic anhydride in the presence of sulfuric acid. The initially obtained triacetate, also used for the manufacture of rayon fibers and films and sheets, is hydrolyzed to yield cellulose acetate. Ethyl cellulose is made by reacting cellulose with sodium hydroxide and ethyl chloride. The plastic grade materials contain 44.5 to 48 % ethoxyl groups.

Ethyl cellulose is tough and moderately flexible, and it is used for flashlight cases, fire extinguisher components, and electrical parts. Cellulose nitrate with a nitrogen content of 11 to 13 % is used for fashion acessories and decorative inlays. Cellulose triacetate is used to produce films and sheeting by solvent casting. These films are used in the production of visual aids, graphic art displays, photographic albums, and protective folders. They are also used in electrical applications and as a base for photographic films. Cellulose acetate contains 38 to 40 % acetyl groups, and the propionates and the butyrates are tougher than the acetate. Cellulose acetate is widely used to produce extruded tape, tool handles, and electrical appliance housings. Cellulose butyrate is used for outdoor signs. Cellulose copolymers are also produced. For example, cellulose acetate-propionate (39 to 47 % propionyl, 2 to 9 % acetyl groups) is used in blister packages and formed containers. Cellulose acetate-butyrate (26 to 39 % butyryl, 12 to 15 % acetyl groups) is used for curtain walls, recreational vehicle windows, small weather shelters, skylights, and toys.

Cornstarch is used as an additive in the formulation of biodegradable polymers for uses in grocery bags, trash bags, compost bags, mulch films, and disposable diapers. In these formulations a few percent of cornstarch, an autooxidant (to promote oxidative degradation), and light-sensitive agents are also added. A biodegradable polyester, poly (3-hydroxybutyrate), is produced by ICI by fermentation of sugar or ethanol using *Alcaligenes eutrophus*. The cell wall of this microbe contains up to 80 % poly (3-hydroxybutyrate), which can be obtained by extraction. The homopolymer can be used in the fabrication of biodegradable products. Under modified conditions, the same microbe produces a copolymer (PHBV) consisting of 3-hydroxybutyrate and 3-hydroxyvalerate units (see Figure 40).

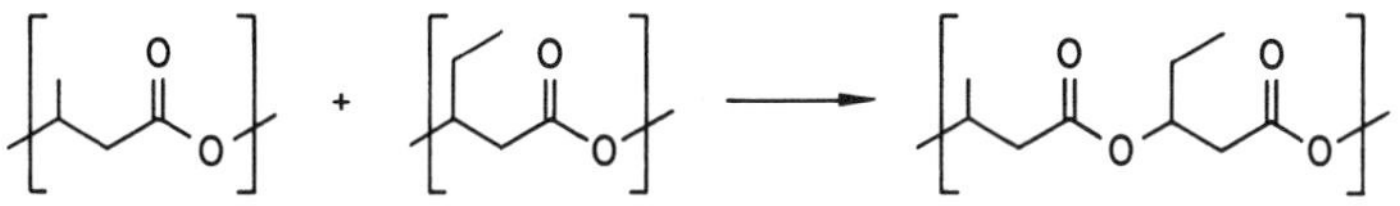

Figure 40 Copolymer derived from 3-hydroxybutyric acid and 3-hydroxyvaleric acid (PHBV).

The copolymer PHBV is more flexible. It has a lower melting point compard to the homopolymer, which improves its processing. PHBV processes somewhat like polypropylene; that is, it can be injection molded and extruded in conventional equipment.

Biodegradation of these polyesters is caused by common bacteria, fungi, and algae. In aerobic sewage, 100 % degradation of 1 mm thick molded sheets occurs in 6 weeks. In anaerobic sewage and soil at 25 °C, complete degradation takes 60 to 75 weeks. In seawater at 15 °C, about 350 weeks is required for complete degradation. The biodegradable polyesters are intended for packaging, for slow-release encapsulation, and for the fabrication of implants that will be completely absorbed by the body. The biodegradable thermoplastic polyester can also be produced in transgenic plants, which can lower its manufacturing cost.

Recently it was observed that bacteria living on longer chain hydrocarbons produce elastomeric polymers with rubberlike properties.

Unsaturated Polyesters

Monomers:	Phthalic anhydride, maleic anhydride, fumaric acid, isophthalic acid, ethylene glycol, propylene glycol, diethylene glycol, styrene
Polymerization:	Bulk polycondensation followed by free radical initiated vinyl polymerization
Major Uses:	Construction, marine applications, automotive applications
Major Producers:	American Cyanamid (Cyglas), Atlas (Atlac), BASF (Palatal), Bayer (Leguval), Hoechst (Alpolit), Koppers (Dion), Montedison (Gabraster), Occidental (Durez), Reichhold (Polylite), Rhône-Poulenc (Stratyl)

Unsaturated polyester matrix materials are produced by first reacting the acid/glycol mixture to give oligomers with molecular weights of 1500 to 3000. To reduce the viscosity, these polyesters are dissolved in styrene monomer (30 to 50 % concentration). Addition of fibers and curing (2 min at 150 °C) with peroxide initiators produces the final network polymer consisting of the original polyester oligomers, which are now interconnected with polystyrene chains. Phthalic anhydride derived polymers have a short cycle time. In contrast, the more corrosion-resistant polyesters made from isophthalic acid require an extended "cooking time" to develop maximum chemical resistance. For automotive applications, appearance is of importance. The unsaturated polyesters used in this application have to perform well with low profile additives and shrink-control agents, and propylene glycol maleate resins are preferred.

In the so-called SMC process, the mixture of resin, fibers, and fillers is held between sheets of polyethylene film until it thickens to a leathery sheet molding compound. Thickening is achieved by reacting free carboxyl end groups with magnesium oxide. This curing or ripening of the matrix converts the soft, sticky mass to a handleable sheet, which takes usually a day or two. A typical formulation consists of 30 % chopped glass fiber, 30 % ground limestone, and resin. The cured sheets or slugs are molded under pressure to give glass fiber reinforced plastic parts. Approximately 80 % of all polyesters are used in reinforced products.

Major markets for reinforced polyesters are appliances, business equipment, construction, and corrosin-resistant products, and in electrical, marine, and transportation applications. In transportation, reinforced plastics are used to achieve weight reduction. In marine applications, pleasure boats as well as larger commercial vessels are being constructed with SMC hulls, which are obtained by the "layer-up" or "spray-up" process. The spray-up process uses gel coat resins based on isophthalic acid and neopentyl glycol. Also, construction industry tubs and shower stalls are increasingly fabricated from glass-reinforced unsaturated polyesters.

Solid alkyd resins, which do not contain styrene monomer, are used in the production of electrical grade molding compounds. The total use of unsaturated polyesters in the United States in 1990 was 600,000 tons.

Aromatic Polyesters

Monomers:	p-Hydroxybenzoic acid, terephthalic acid, isophthalic acid, bisphenol A
Polymerization:	Bulk polycondensation or interfacial polycondensation
Major Uses:	High performance engineering thermoplastics, abradable seals, plasma coatings
Major Producers:	Amoco (Ardel), Carborundum (Ekonol), Du Pont (Arylon, Bexloy M), Hoechst Celanese (Durel)

Amorphous aromatic polyesters are used as engineering plastics. The AABB-type polymers are obtained fom bisphenol A and diphenyl isophthalate or isophthaloyl chloride, and the diacetate of bisphenol A can also be used in conjunction with the acids. In 1989 researchers form G.E. demonstrated that macrocyclic oligomers could be obtained from bisphenol A and isophthaloyl chloride using the interfacial polymerization technology. However, in this reaction only 50 % of the oligomers and 50 % of polyarylates were obtained. The AB-type polyarylates are obtained by homopolymerization of the phenyl ester of p-hydroxybenzoic acid. To obtain amorphous polyarylates, mixtures of isophthalic and terephthalic acid derivatives are used.

Polyarylates have excellent thermal stability, a good overall balance of strength and toughness over a wide temperature range, and good high temperature creep resistance and flexural recovery. Their processing character is similar to polycarbonates. The heat deflection temperature of the AABB-type polymers is about 174 °C, their T_g's are about 190 °C, and the flammability characteristics and electrical properties are excellent. The homopolymer derived from *p*-hydroxybenzoic acid has been marketed by Carborundum under the trade name Ekonol. This material is used in plasma coatings. The melt temperature of Ekonol is above 500 °C, and therefore it can be processed only by sintering in hot presses.

Thermoplastic polyarylates are used in automotive and electrical/electronic applications, in small appliances, and in business machines.

Also copolymers of polyarylates and polycarbonates are produced by polycarbonate manufacturers using the interfacial polycondensation of the sodium salt of bisphenol A with a mixture of phosgene and aroyl chlorides. Examples are Bayer's APE, Dow's AEC, and G.E.'s Lexan PPC. These copolymers are also used in automotive applications.

Polycarbonate (PC)

$$\left[-O-\underset{}{\bigcirc}-\underset{CH_3}{\underset{|}{\overset{CH_3}{\overset{|}{C}}}}-\underset{}{\bigcirc}-O-\underset{\underset{O}{\|}}{C} \right]_n$$

Monomers:	Bisphenol A, phosgene
Polymerization:	Interfacial polycondensation
Major Uses:	Glazing (25 %), transportation (15 %), communications and electronics (12 %), industrial (11 %), sporting equipment (10 %)
Major Producers:	Anic (Sinvet), Atochem (Orgalan), Bayer (Makrolon), Dow (Calibre), G. E. (Lexan)

The transparent amorphous polycarbonates are produced by interfacial polycondensation of the sodium salt of bisphenol A and phosgene. this reaction is conducted in methylene chloride/water.

$$HO-\underset{}{\bigcirc}-\underset{\underset{CH_3}{|}}{\overset{\overset{CH_3}{|}}{C}}-\underset{}{\bigcirc}-OH + Cl-\underset{\underset{O}{\|}}{C}-Cl \xrightarrow[H_2O, NaOH]{CH_2Cl_2}$$

$$\left[-O-\underset{}{\bigcirc}-\underset{\underset{CH_3}{|}}{\overset{\overset{CH_3}{|}}{C}}-\underset{}{\bigcirc}-O-\underset{\underset{O}{\|}}{C} \right]_n + 2NaCl$$

Polycarbonates can also be obtained by solution polymerization in pyridine, where the solvent acts as the hydrogen chloride scavenger. A third approach is the transesterification reaction of phenyl carbonate with bisphenol A. This reaction can conceivably be conducted continuously in a vented extruder, the by-product being phenol:

$$HO-\phi-C(CH_3)_2-\phi-OH + \phi-O-CO-O-\phi \longrightarrow$$

$$[-O-\phi-C(CH_3)_2-\phi-O-CO-]_n + 2\,\phi-OH$$

In 1989 G.E. scientists announced a new process for polycarbonate based on ring-opening polymerization of cyclic oligomers. The cyclic oligomers were made by interfacial polymerization of the bischloroformate derived from bisphenol A with bisphenol A in the presence of triethylamine and sodium hydroxide using methylene chloride as the organic phase. In this manner 85 % of macrocyclic carbonates ($n = 2$ to 20), and 15 % of polycarbonate was obtained. The PC can be removed by precipitation with acetone.

$$ClCOO-\phi-C(CH_3)_2-\phi-OCOCl \xrightarrow[Et_3N, NaOH]{bis\ A} \text{(cyclic oligomer)}_n \xrightarrow{cat.} PC$$

The cyclic oligomers were polymerized in solution (*o*-dichlorobenzene) or in the melt at 250 °C, using anionic catalysts, such as lithium trifluoroethoxide, or bis(acetyl acetonato)titanium diisopropoxide. In this manner polycarbonates are produced with molecular weights ranging from 70,000 to 700.000 (in the reaction of phosgene with bisphenol A, molecular weights ranging from 40,000 to 60,000 were obtained). The feasibility of melt polymerization indicates that the macrocyclic oligomers can be used in reaction injection molding (RIM) applications. Fabrication of parts from the oligomers allows glass loadings of up to 70 %.

Polycarbonates possess exceptionally high impact strength, combined with good electrical properties, thermal stability (T_g 140 °C), and creep resistance. Use temperatures range from –51 to 132 °C at 264 psi, the materials's heat deflection temperature. Additives such as glass fiber reinforcement can be used, and coatings are sometimes used on polycarbonate sheet to improve mar and chemical resistance. Fire retardant grades of polycarbonates are produced using tetrabromobisphenol A as a comonomer.

A new class of co-polycarbonates were introduced by Bayer under the trade name Apec HT. These co-polycarbonates are based on bisphenol-A and bis-phenol TMC (derived from hydrogenated isophorone). The new co-polycarbonates have better melt viscosities and Tg's of up to 205 °C.

Polycarbonates are water clear and transparent, having roughly 89 % transmittance and less than 1 % haze. Polycarbonate is the primary matrial of choice for the molding of compact discs (CD) and electronic memory discs. Other major applications include glazing, lighting, transportation, appliances, signs, sports equipment, returnable bottles, and solar collectors. Double-wall extruded polycarbonate is used in greenhouse glazing applications. Football, motorcycle, and snowmobile helmets are examples of the use of PCs in sports equipment. The worldwide polycarbonate consumption in 1991 was 600,000 tons.

15 Polyamides

Introduction

The commercial development of polyamides started in the early 1930s when Hill, a coworker of Carothers, stuck a glass rod into the melt of a polyamide, made by melt condensation of the salt of a dicarboxylic acid and a diamine (AABB-type polycondensation), and pulled some fibers. In 1938 Du Pont started the manufacture of nylon-6,6. In a parallel development, Schlack in Germany developed AB-type polyamides by homopolymerization of cyclic lactams, and nylon-6, derived from caprolactam, was introduced in 1939. New meltspinning processes were developed, and the golden age of synthetic fibers began.

Nylon's first major business target was the silk hosiery market. After overwhelming acceptance of nylon hosiery, the tricot and home furnishing markets were penetrated mainly because of nylon's excellent knittability and durability. Industrial applications, especially in carpeting, followed. Today both nylon-6,6 and nylon-6 are still the leading products, comprising about 90 % of the nylon market. Major end uses of nylon fibers are in home furnishings (61 %), apparel (18 %), and tire cord (11 %).

The new aromatic nylons, commercially introduced in 1961, have expanded the maximum use temperatures to well above 200 °C, and the new high tenacity, high modulus aramides have provided new levels of properties ideally suited for tire reinforcement. New families of nylon molding compounds have been introduced, and the domination of the nylon market by the classical six-carbon derived products is challenged by the many new products introduced in the 1970s and 1980s.

Aliphatic Polyamides

$$[-NH(CH_2)_6NH-\underset{O}{\underset{\|}{C}}-(CH_2)_4-\underset{O}{\underset{\|}{C}}-]_n \qquad [-NH(CH_2)_5-\underset{O}{\underset{\|}{C}}-]_n$$

Nylon 6,6   Nylon 6

Monomers:	Adipic acid, hexamethylenediamine, caprolactam
Polymerization:	Bulk polycondensation
Major Uses:	Home furnishings (61 %), apparel (18 %), tire cord (11 %)

Major Producers: Nylon-6: Akzo (Akulon), Allied (Capron), Atochem (Orgamide), BASF (Ultramid B), Bayer (Durethan B), Ems Chemie (Grilon), Hoechst (Fosta), ICI (Maranyl F), Montedison (Rhenyl), Rhône-Poulenc (Technyl), Snia (Sniamid)
Nylon-6,6: Akzo (Akulon), BASF (Ultramid A), Bayer (Durethan A), Du Pont (Zytel, Minlon), ICI (Maranyl A, Verton), Monsanto (Vydyne), Montefibre (Nailonplast A), Snia (Sniavitrid)
Nylon-11: Atochem (Rilsan 11)
Nylon-12: Atochem (Rilsan 12), Huels (Vestamid)

The commercial success of nylon-6 and nylon-6,6 is due to outstanding properties and an economically attractive raw material base. Nylon-6,6 is produced by melt condensation of adipic acid with hexamethylenediamine. Both monomers contain six carbon atoms (nylon-6,6) and therfore are obtainable from either benzene or other petrochemical products. Adipic acid is produced by air oxidation of cyclohexane to give a mixture of cyclohexanone and cyclohexanol, followed by further oxidation with nitric acid (see Figure 41). Hexamethylenediamine is produced by the reduction of adiponitrile, obtained from butadiene and hydrogen cyanide, and by the oxidative coupling of acrylonitrile. However, not all caprolactam is made from cyclohexane; phenol accounts for nearly half of the caprolactam produced in the United States.

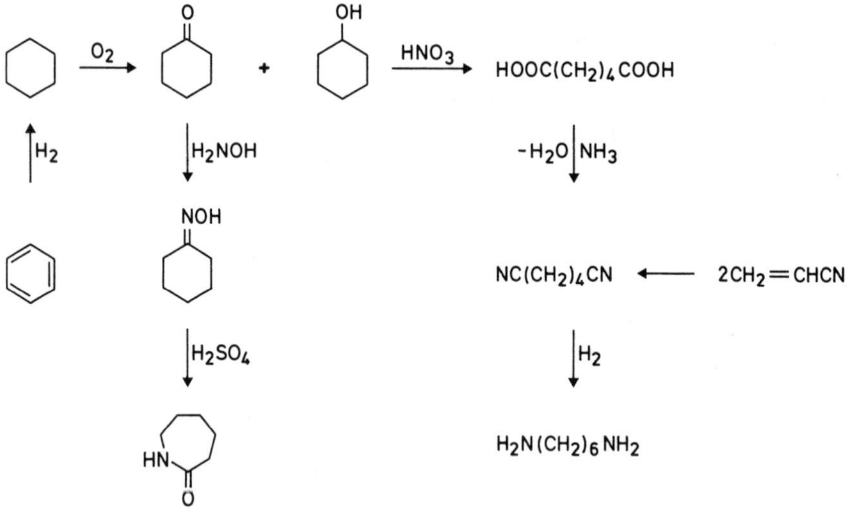

Figure 41 Basic raw materials for nylon-6 and nylon-6,6

Two basic types of condensation polymer are possible. Nylon-6,6 is the typical example of the polycondensation of two difunctional monomers, adipic acid and

15 Polyamides

hexamethylenediamine. This type of polycondensation is generally referred to as the AABB reaction. In contrast, polycondensation of one difunctional monomer is referred to as an AB reaction. The AABB-type polycondensation has the advantage that the stoichiometry is fixed by salt formation. In contrast, AB polyamides are usually prepared by ring-opening polymerization of a cyclic lactam. This reaction requires only a catalytic amount of water, because the water required for ring opening is replenished by the simultaneously occurring polycondensation:

$$\text{caprolactam} \xrightarrow{H_2O} [H_2N(CH_2)_5COOH] \xrightarrow{-H_2O} [-HN(CH_2)_5-C(=O)-]_n$$

In addition to the well-known nylon-6,6, several other AABB products, such as nylon-4,6, nylon 6-10, nylon-6,12, and nylon-12,12, have been introduced. The 1,4-diaminobutane used in nylon-4,6 is made from acrylonitrile and hydrogen cyanide, followed by hydrogenation. The highly crystalline nylon-4,6, produced by DSM (Stanyl), has improved thermal deformation characteristics under high loadings, which improves its temperature creep resistance over nylon-6,6.

Polyamides are highly crystalline hydrogen-bonded linear polymers. The difference in the number of carbon atoms between the amide groups results in significant differences in mechanical and physical properties of the resultant polymers. The lower the number of carbon atoms, the higher the specific gravity, the melting point, and the mechanical properties, but the higher the moisture absorption in a humid environment, as well.

The melt temperatures of some AABB-type polyamides are shown in Table 36. The higher the number of carbon atoms, the lower the melt temperature. The difference in melting points is exploited in bicomponent fibers made by spinning both polymers simultaneously through the same orifice under conditions ensuring that the polymer streams do not mix.

Table 36 Melt temperatures of AABB Polyamides

Polymer	T_m (°C)
4,6	295
6,6	265
6,8	240
6,9	205
6,10	225
6,12	217

The melt temperatures of the AB polyamides are listed in Table 37. The homopolymers below nylon-6 are hydrophilic, while homopolymers above nylon-6 are more hydrophobic. In addition to nylon-6, only nylon-11 and nylon-12 have found applications, mainly because of their lower moisture absorption. Nylon-11 is made by Atochem (Rilsan 11).

The monomer, amino undecanoic acid, is obtained from castor oil by a number of chemical transformations. Nylon-12 is manufactured by ring-opening polymerization of laurolactam. Cyclotrimerization of butadiene produces cyclododecatriene, which is converted in several steps to laurolactam, the starting material for nylon-12. Nylon-11 was introduced by Atochem in the 1960s, and nylon-12 was developed in the 1970s. Major uses for nylon-11 and nylon-12 are in truck and automotive tubing and hoses and in powder coatings.

Table 37   Melt Temperatures of AB Polyamides

Polymer	T_m (°C)
Nylon-4	265
Nylon-6	215
Nylon-7	223
Nylon-11	194
Nylon-12	179

The absorption of moisture in nylon molding compounds results in poor dimensional stability and deterioration of mechanical and electrical properties (see Table 38).

Table 38   Absorption of Moisture in Nylons

Polymer	Water Absorption* at 20 °C	Dimensional Change* (%)
6	2.7	0.7
6.6	2.5	0.6
11	0.8	0.12

* AT 50 % relative humidity.

A special polyamide fiber with silklike properties was developed by Du Pont (Qiana). The monomers for Qiana fibers are dodecandioic acid and bis (4-aminocyclohexyl) methane.

$$HOOC(CH_2)_{10}COOH + H_2N-\langle\rangle\langle\rangle-NH_2 \longrightarrow$$

$$-[-\underset{O}{\overset{\|}{C}}-(OH_2)_{10}-\underset{O}{\overset{\|}{C}}-NH-\langle\rangle\langle\rangle-NH-]_n$$

The cycloaliphatic diamine is obtained by hydrogenation of the corresponding aromatic diamine, and a mixture of stereoisomers is obtained in this reaction. A mixture enriched in the trans-trans isomer is used in the production of the polyamide. Qiana was tailored to have silklike properties, but the melting characteristics also differ from nylon-6,6:

	T_m (°C)	T_g (°C)
Qiana	205	135
Nylon-6,6	265	90

The high glass transition temperature of Qiana fibers assures that the polymer will remain in the glassy state during fabric laundering, resisting wrinkles and creases. Qiana fibers are no longer produced by Du Pont.

Aliphatic-Aromatic Polyamides

Monomers:	Isophthalic acid, terephthalic acid, hexamethylenediamine, bis(4-aminocyclohexyl) methane, trimethylhexamethylenediamine
Polymerization:	Bulk polycondensation
Major Uses:	Tubings for fiber optics, automotive, filter bowl units, milk handling equipment
Major Producers:	BASF (Ultramid KR), Bayer (Durethan C, T), Du Pont (Zytel 330), Dynamit Nobel (Trogamid T), Ems Chemie (Grilamid TR 55)

The aliphatic-aromatic polyamides are amorphous and transparent polymers with good resistance to a wide range of solvents, chemicals, oils, and greases. Amorphous nylons were introduced by Dynamit Nobel in 1967. The amorphous character of the polymers is achieved by using mixtures of comonomers. Du Pont is using mixtures of isophthalic and terephthalic acids and hexamethylenediamine and bis(4-aminocyclohexyl)methane (> 50 % trans-trans isomer). Dynamit Nobel obtained Trogamid T by a polycondensation of dimethyl terephthalate with trimethylhexamethylenediamine (50/50 mixture of 2,2,4- and 2,4,4-trimethyl isomers). Ems Chemie is utilizing a mixture of laurolactam (nylon-12 monomer) and bis (4-amino-3-methylcyclohexyl) methane with isophthalic acid in the manufacture of its amorphous nylons.

Aromatic Polyamides

Monomers:	Isophthalic acid, terephthalic acid, phenylenediamines
Polymerization:	Interfacial polycondensation using acid chlorides
Major Uses:	Ballistic vests, heat-resistant clothing, tire reinforcement, smoke stack filtration
Major Producers:	Akzo (Twaron), Du Pont (Nomex, Kevlar), Teijin (Conex)

Aromatic polyamides are produced mainly because of their superior heat resistance. In 1961 Du Pont introduced heat-resistant Nomex fibers, which are produced from isophthaloyl chloride and *m*-phenylenediamine. Kevlar, made from terephthaloyl chloride and *p*-phenylenediamine, was introduced by Du Pont in 1973.

Nomex

(reaction scheme: isophthaloyl chloride + m-phenylenediamine → Nomex)

Kevlar

(reaction scheme: terephthaloyl chloride + p-phenylenediamine → Kevlar)

Kevlar is ideally suited for tires and ballistic vests because of its strength. It is as strong as steel at one-fifth the weight. The Kevlar fibers have a T_g above 300 °C and can be heated without decomposition to temperatures exceeding 500 °C. The dimensional stability is outstanding, with essentially no creep or shrinkage as high as 200 °C.

In view of the high melt temperatures of these polymers (see Table 39) and their poor solubility in conventional solvents, special techniques were required to produce the fibers. Aramides are spun from highly polar solvents, such as N-methylpyrrolidone/CaCl$_2$.

Table 39 Thermal Stability of Aramides

Dibasic Acid	Diamine	T_m (°C)
Terephthalic acid	p-Phenylenediamine	(500)*
	m-Phenylenediamine	> 400
Isophthalic acid	p-Phenylenediamine	> 400
	m-Phenylenediamine	365**

* Kevlar;   ** Nomex.

The properties of synthetic fibers range from low modulus, high elongation fibers like Lycra (see Chapter 12) to high modulus, high tenacity fibers such as Kevlar. In between, almost any combination of properties can be built into macromolecules. The search for still better performance will continue. While the maximum tenacity value for nylon-6,6 is 215 g/denier, we actually only produce nylon-6,6 fibers with a tenacity value of 10 g/denier. A breakthrough in fiber strength and stiffness has been achieved with Kevlar and graphite fibers. These new fibers have created superior composite materials, generally referred to as fiber-reinforced plastics (FRPs).

Low density, high specific strength (strength/density), and high stiffness (modulus/density) make aramides promising as metal replacement materials. Glass fiber reinforced boats and other recreational vehicles are already a reality, and the automotive industry is investigating the use of glass fiber reinforced polyurethanes for fender, hood, trunk, and door panel applications. Today, fishing rods are filament wound with graphite and Kevlar fibers, and fiber reinforced golf club shafts, tennis rackets, skis, ship masts, and other products are being tested in the marketplace. Significant quantities of advanced graphite and graphite/Kevlar composites are used in Boeing's 757 and 767 aircraft. The global consumption of aramides in 1990 was 18,000 tons. It is anticipated that the use of FRPs will grow rapidly in the years to come. The use of advanced composites on the Boeing 767 is shown in Figure 42.

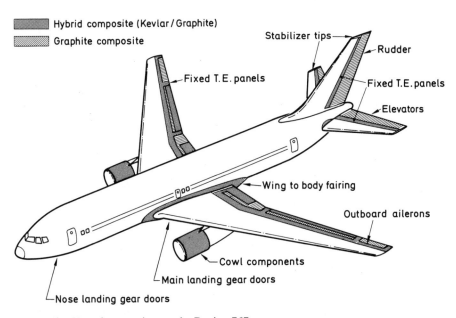

Figure 42 Use of composites on the Boeing 767

Fiber-reinforced polymer composites, such as hybrid graphite fiber composite systems, achieve dramatic weight savings (up to 70 %). Properties of today's high performance fibers are summarized in Table 40.

Table 40 Fiber Properties

Fiber	Density	Tensile Strength	Modulus	Cost ($/lb)
E-Glass	2.63	2 415	72	0.50
S-Glass	2.63	3 450	90	2.30
Kevlar	1.45	2 760	130	9.00
Graphite I	1.75	2 760	235	20–32
Graphite II	1.85	2 415	220	75.00

Boron, alumina, and silicon carbide fibers are also high performance fibers, but they are too expensive for commercial applications. Coatings are often applied to the fibers prior to mixing with the matrix polymer. The wetting of the fibers can be enhanced by spreading a thin layer of the matrix polymer on the fibers. Practically all glass fibers for composites are treated with silanes, such as trichlorovinylsilane, in which the end groups form Si-O bonds with the glass surface, and the organic group is compatible with the matrix polymer.

Continuous FRPs are highly anisotropic. If all fibers are aligned in one direction, maximum properties are achieved in that direction. Properties decrease rapidly in the direction away from the fiber direction. To generate more isotropic properties, alternate layers of fibers are frequently employed.

Polyamide Imides

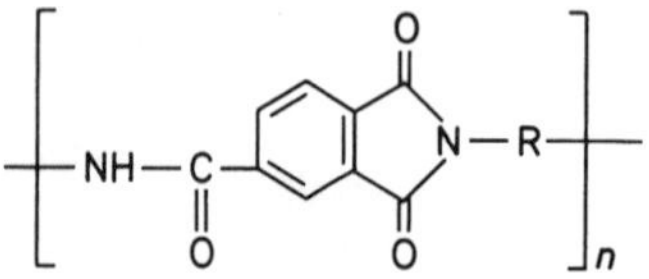

Amide Imide

Monomer:	Trimellitic acid anhydride, aromatic diamines (MDA)
Polymerization:	Two-step or one-step polycondensation
Major Uses:	Film, wire enamel, molded parts
Major Producer:	Amoco (Torlon)

Polyamide imides (PAIs) are amorphous, high temperature engineering thermoplastics used in the aerospace, transportation, chemical processing, and electronic industries. The PAI polymers are produced from tricarboxylic acid anhydrides, such as trimellitic acid anhydride, and aromatic diamines, such as MDA. Amoco introduced its Torlon PAI in the early 1970s. The polycondensation can be conducted in one step or in two steps (Figure 43). In the two-step reaction a solution of a polyamide acid is initially formed, and the final polymer can be generated in a baking process. This approach is being used for the formultation of heat-resistant wire and cable coatings.

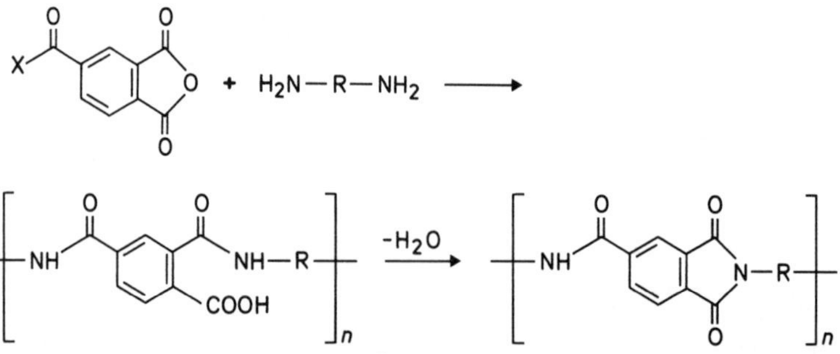

Figure 43 Synthesis of polyamide imides

A different synthetic approach to polyamide imides was pioneered by Upjohn. Instead of the aromatic diamine, the corresponding diisocyanate is used. For example, reaction of trimellitic acid anhydride with 4,4'-diisocyanatodiphenylmethane (MDI) in a polar solvent affords the polyamide imide and carbon dioxide gas. In this reaction it is not necessary to convert the carboxylic acid group into an acid chloride.

Polyimides

Monomers:	Pyromellitic acid dianhydride, benzophenonetetracarboxylic dianhydride, aromatic diamines, bis (maleimide), MDI
Polymerization:	Solvent polycondensation
Major Uses:	Film, molding compounds, fibers
Major Producers:	Du Pont (Kapton, Vespel), G.E. (Ultem), Lenzing (P 84), Rhône-Poulenc (Keremid, Kinel)

Polyimides are produced from tetracarboxylic acid dianhydrides and aromatic diamines or aromatic diisocyanates. Polyimides are thermally stable and retain a significant portion of their physical strength at temperatures up to 482 °C (900 °F) in short-term exposure. For prolonged exposures they can be used at about 260 °C (500 °F). In addition to thermal and oxidative stability, the polyimides have excellent electrical properties.

Polyimides based on pyromellitic dianhydride and m-phenylenediamine were pioneered by Du Pont. These polymers could be processed only by sinter molding.

In 1969 Rhône-Poulenc introduced lower performance polyimides based on MDA-derived bis(maleimides), which were subsequently "cured" by an addition reaction with more of an aromatic diamine. Polyimides based on the reaction of benzophenone tetracarboxylic acid dianhydride with 4,4'-diisocyanatodiphenylmethane (MDI) were introduced in the early 1970s. These polymers were later used by Lenzing in the production of polyimide fibers.

In 1982 G.E. introduced the first thermoplastic polyimide (Ultem). This amorphous high temperature thermoplastic polyimide is based on the reaction of an ether dianhydride, made from bisphenol A and nitrophthalic anhydride, with m-phenylenediamine.

Ultem has a continuous use temperature of about 170 to 180 °C, and it is applied in many applications requiring ease of processing and thermal stability. Examples include jet engine parts, aircraft cabin components, fuel system components, printed circuit boards, and industrial appliances.

16 Formaldehyde Resins

Introduction

The phenol-formaldehyde and urea-formaldehyde resins are the most widely used thermoset polymers. Thermosets are highly crosslinked materials obtained by stepwise polymerization processes. The crosslinking process in usually referred to as curing. Often, lower molecular weight prepolymers are used as precursors, and the final form and shape are generated under heat and pressure. In this step water is generated, usually in the form of steam because of the high processing temperatures. Fillers, such as mineral and glass fibers, can be added to improve the physical properties of the molded part. The preferred form of processing is injection molding, but compression molding and transfer molding are also used.

The high volume thermosets owe their existence to the relatively low cost of the components, and to their superior thermal and chemical resistance. The phenolic resins introduced by Baekeland in 1907, are mainly used as binder resins for plywood and foundry core applications, but significant amounts are used in molding applications.

Phenol-Formaldehyde Resins (PF)

Monomers:	Phenol, formaldehyde
Polymerization:	Base- and acid-catalyzed stepwise polycondensation
Major Uses:	Plywood adhesive (34 %), glass fiber insulation (19 %), molding compound (8%)
Major Producers:	Bakelite (Bakelite), Borden, Dynamit Nobel (Trolitan), Georgia Pacific, Occidental (Durez), Plastics Engineering (Plenco), Rhône-Poulenc (Progilite)

The polymerization of phenol with formaldehyde is a polycondensation process that can be catalyzed by alkali or acids. The original Bakelite materials are base catalyzed, while the novolacs are polymers obtained by acid catalysis. Phenolic insulation foams are also produced using acid catalysis.

In the base-catalyzed condensation of phenol with formaldehyde, low molecular weight liquid prepolymers, called resoles, are produced using an excess of formaldehyde. These prepolymers have reactive hydroxymethyl groups attached to the nuclei, and their shelf life is limited. The curing mechanism for resoles is shown in Figure 44.

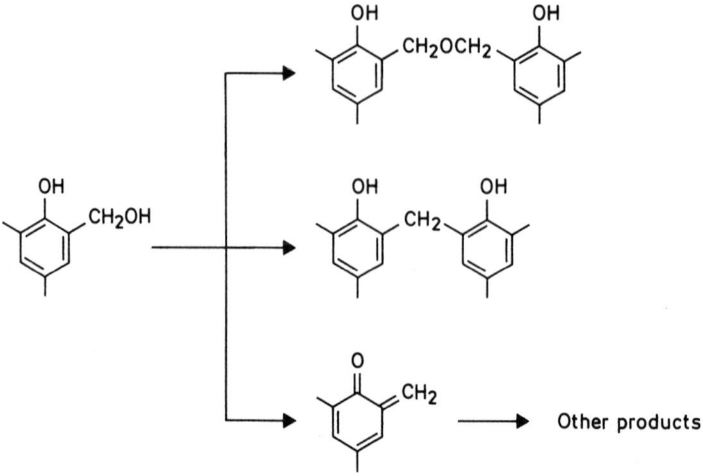

Figure 44 Curing mechanism for resoles

The formation of quinonemethide groups, as depicted on the bottom of Figure 44, is responsible for discoloration as well as for crosslinking by cycloaddition rather than condensation. Resoles are used as binder resins for plywood, but they are also used in the formation of phenolic insulation foams.

Formation of novolacs is an acid-catalyzed reaction leading to carbon-carbon bond formation as follows:

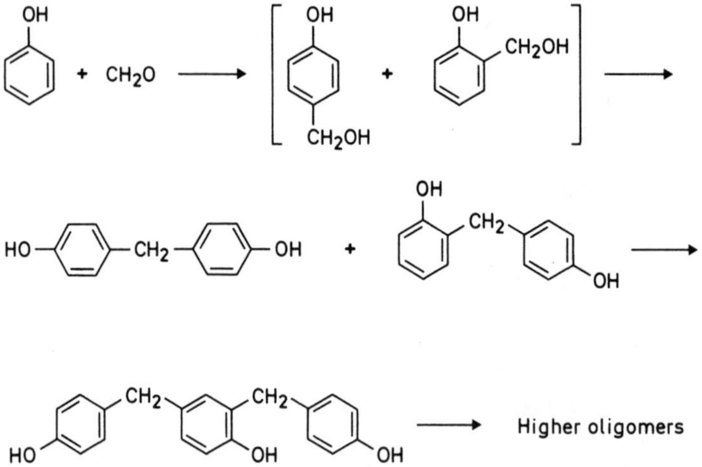

This reaction proceeds best with a slight deficiency of formaldehyde. Final crosslinking is achieved by further reaction with paraformaldehyde or hexamethylene-tetramine (hexacure) in the final fabrication step.

Because of their superior dimensional stability and electrical properties, molded parts are used in distributor caps, fuse boxes, and other electrical outlets.

Resorcinol instead of phenol is used for specialty phenolics because it increases the curing speed. It is used in water-soluble resole resins for tire cord adhesives. Another specialty phenolic resin is made from aralkyl ethers, such as bismethoxy-*p*-xylene, and phenol, unsing either hexamethylenetetramine or selected epoxy compounds as curing agents. Phenol-aralkyl ether resins find uses in many high perfomance electrical and structural applications.

Urea-Formaldehyde Resins (UF)

Monomers:	Urea, fomaldehyde
Polymerization:	Stepwise polycondensation
Major Uses:	Particle board binder resins (60 %), paper and textile treatment (10 %), molding compounds (9 %), coatings (7 %)
Major Producers:	American Cyanamid, Borden, Georgia Pacific, Hercules, Monsanto

The amino resins or aminoplasts are produced by the condensation of urea, diamines, or melamine with formaldehyde. Formaldehyde reacts with the amino groups to give amino methylol derivatives, which undergo further condensation with free amino groups to produce resinous products. In contrast to phenolic resins, products derived from urea and melamine are colorless.

Urea-formaldehyde resins were developed in 1920 as adhesives, casting compositions, and materials for textile treatment. The basic reactions involved in the formation of the crosslinked thermoset products are shown in Figure 45.

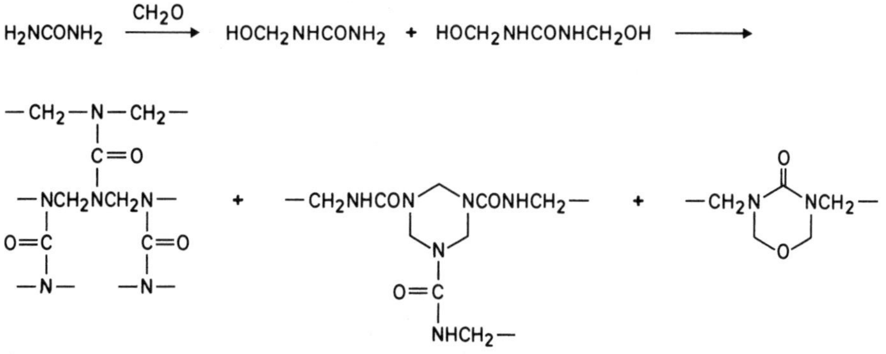
H2NCONH2
CH2O
HOCH2NHCONH2 + HOCH2NHCONHCH2OH
—CH2—N—CH2—
C=O
—NCH2NCH2N—
O=C
C=O
—N—
—N—
+
—CH2NHCON
NCONHCH2—
N
O=C
NHCH2—
+
O
—CH2N
NCH2—
O

Figure 45 Urea-formaldehyde condensation products

The reaction product of urea with glyoxal and 2 moles of formaldehyde (DMDHEU) is used as a wrinkle recovery, wash-and-wear, durable-press agent in the textile industry.

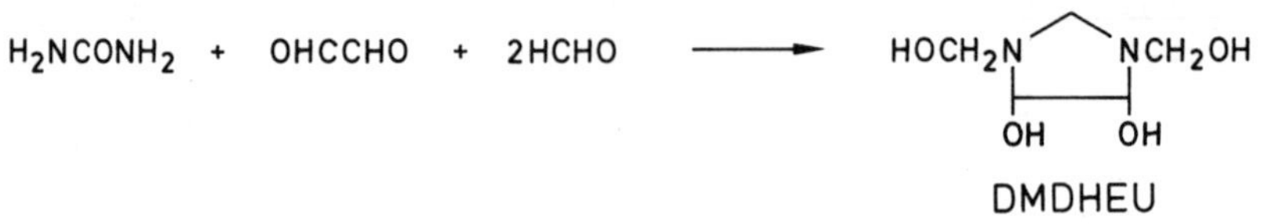
H_2NCONH_2 + OHCCHO + 2HCHO
$HOCH_2N$
NCH_2OH
OH OH
DMDHEU

The textiles are treated with a solution containing the chemical, and heating bonds the material to celullose or polymerizes the chemical into the fabric surface.

Melamine-Formaldehyde Resin (MF)

Monomers:	Melamine, formaldehyde
Polymerization:	Stepwise polycondensation
Major Uses:	Dinnerware, table tops, coatings
Major Producers:	American Cyanamid, Borden, Georgia Pacific

Melamine, 2,4,6-triamino-1,3,5-triazine, reacts with fromaldehyde to form a water-soluble A-stage resin. Evaporation and drying produces a crosslinked B-stage resin, which is then ground to give a powder that is used as molding resin for a variety of mechanical parts or household goods, and as laminating resin for counter, cabinet, and table tops. This material is commonly known as Formica. The watersoluble A-stage resin is also used for textile treatment to impart crease resistance, stiffness, shrinkage control, water repellency, and fire resistance. The synthesis of Formica is shown in Figure 46.

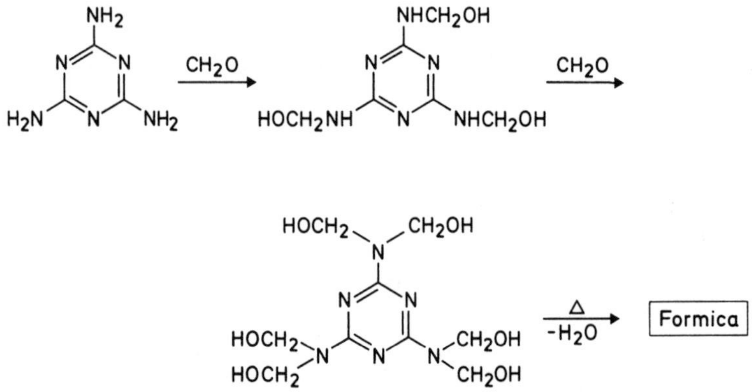
NH_2
H_2N
NH_2
CH_2O
$NHCH_2OH$
$HOCH_2NH$
$NHCH_2OH$
CH_2O
$HOCH_2$
CH_2OH
$HOCH_2$
$HOCH_2$
CH_2OH
CH_2OH
Δ
$-H_2O$
Formica

Figure 46 Synthesis of Formica

Polymer formation occurs via *N*-methylol compounds. The functionality depends on the ratio of formaldehyde to melamine. Hexamethylolmelamine is the most stable and the easiest to purify. Heating hexamethylolmelamine to 150 °C converts it to an insoluble glassy polymer. The crosslinking mechanism is similar to that observed for urea-formaldehyde resins; triazine and hexahydrotriazine units provide the basic network structure (Figure 45).

Part IV
Special Polymers

17 Heat-Resistant Polymers

Introduction

For polymers to be useful as construction materials, it is necessary to increase the melt and softening temperatures of thermoplastics (linear polymers) or thermosets (crosslinked polymers). The category of the so-called engineering thermoplastics includes such relatively low melting materials as ABS, polyacetal, polycarbonate, and the molding grade nylons. However, the upper limit of use temperatures for these materials is only 80 to 120 °C.

The melting points of polymers are generally increased by polar groups in the polymer backbone or by constructing stereoregular polymers. Decreases in mobility of chains by steric hindrance also increases the melting point. Examples for polyolefins are listed in Table 41. The increase in melt temperature from polyethylene to isotactic polypropylene is approximately 40 °C. The bulky substituent in the stereoregular poly (3,3-dimethyl-1-butene), which has a pendant t-butyl group, increases the melt temperature by 80 °C over polypropylene with a pendant methyl group. The fluorosubstituent in polytetrafluoroethylene stiffens the chains to the point that melting does not occur prior to decomposition.

Table 41 Melt Temperatures of Some Polyolefins

Polymer	T_m (°C)
Polyethylene	140
Isotactic polypropylene	180
Isotactic poly(4-methylpentene)	235 – 240
Isotactic poly(3,3-dimethyl-1-butene)	260
Tetrafluoroethylene	>350

Aliphatic polymer backbones are not preferred for heat-stable polymers, because the weakest bond determines the overall thermal stability of the macromolecules, and the aliphatic carbon – carbon bond has a relatively low bond energy (see Table 42). Oxidation of alkylene groups is also oberserved upon prolonged heating in air.

Table 42 Bond Energies of Common Organic and Inorganic Polymers

Bond	Bond Energy (kcal/mol)	(kJ/mol)
$C_{al}-C_{al}$	83	347
$C_{al}-H$	97	405
$C_{ar}-C_{ar}$	98	410
$C_{ar}-H$	102	426
B–N	105	439
Si–O	106	443
C–F	116	485

Significantly higher use temperatures are achieved with aromatic polymers. For example, poly(p-phenylene) has been synthesized by Marvel, a pioneer in the field of heat-stable polymers, by stereospecific 1,4-cyclopolymerization of cyclohexadiene, followed by dehydrogenation.

This polymer was found to be insoluble and infusible.

Thin films of p-xylylene can be vacuum deposited on substrates. The monomer can be generated only by high temperature laser or plasma induced pyrolysis of p-xylene in the vapor phase, and therefore application is rather limited. This process is reminiscent of the vacuum deposition of metals to the surface of plastic substrates, a practice that is widely used in the automotive industry. The chemistry involved in the formation of poly (p-xylylene) is summarized by the reaction:

Commercial polymers based on the principle of synthesis of polyaromatic compounds include the previously discussed aramide fibers and polyphenylene oxide, arylketones, polyphenylene sulfide, polysulfones, polybenzimidazoles, and liquid crystal polymers.

Another approach to the acievement of stability is in the construction of ladder polymers (Figure 47). If these ladder polymers are aromatized as shown for "black orlon," they are even more stable. Further heating of "black orlon" to 1400 to 1800 °C and simultaneous stretching produces graphite HT. If the heating and stretching are conducted at 2400 to 2500 °C, the high modulus material graphite HM is obtained. Other carbonizable polymers that produce carbon fibers on heating include PVC, PVA, polyvinylidene chloride, cellulose, and phenolic resins. Thermosets generally produce nongraphitizing or glassy carbon.

N-Vinyl-nylon-1

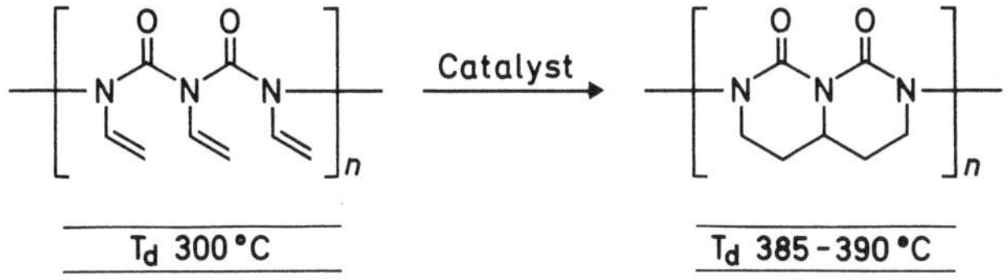

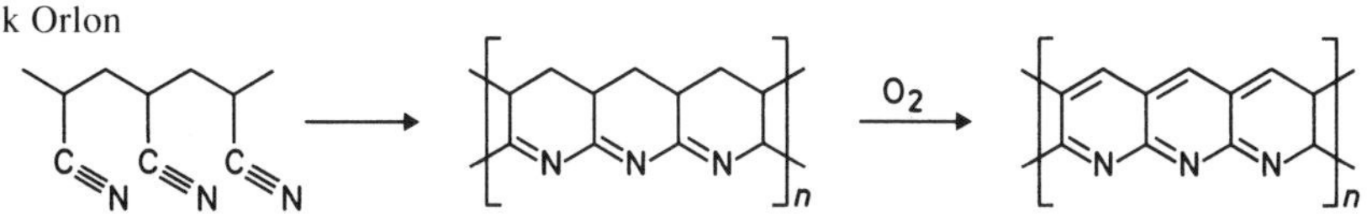

Figure 47 Increase in thermal stability by ring formation (ladder polymers)

Ladder polymers are also produced by polycondensation reactions of tetrafunctional monomers. If a tetrafunctional monomer is reacted with a difunctional monomer as in the formation of polyimides (see Chapter 15), the derived polymer is referred to as a partial ladder or stepladder polymer. If two tetrafunctional monomers are used, as in the formation of polyquinoxaline from an aromatic tetramine and an aromatic tetraketone, such polymer is again referred to as a ladder polymer.

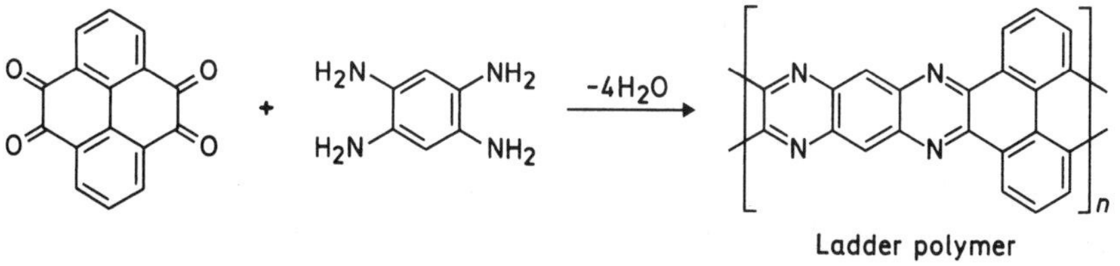

Arylketones

Monomers: 4,4'-Difluorobenzophenone, hydroquinone, 4,4'-dihydroxybenzophenone
Polymerization: Polycondensation
Major Uses: Automotive engine parts, film, wire and cable
Major Producers: BASF (Ultrapek), Hoechst (Hostatec), ICI (Victrex)

Polyether ether ketone (PEEK) is a crystalline high temperature thermoplastic polymer. It is made by condensation polymerization of 4,4'-difluorobenzophenone and the dianion of hydroquinone in molten diphenylsulfone or in a high boiling polar solvent, such as *N*-cyclohexyl-2-pyrrolidone.

The melting point of PEEK is 334 °C, and it has a continuous service temperature of 250 °C. The degree of crystallinity of the polymer can be decreased by producing copolymers with diphenyl units. In this case the comonomer would be 4,4'-dihydroxydiphenyl.

Polyether ketone (PEK) is a partial crystalline polymer made by the condensation polymerization of 4,4'-difluorobenzophenone and the dianion of 4,4'dihydroxybenzophenone, or by self-condensation of 4-fluoro-4'hydroxybenzophenone. It has a high temperature performance similar to that of PEEK.

Applications of the polyarylketones include high performance automotive and industrial applications and electronics (circuit boards).

Polyphenylene Sulfide (PPS)

Monomers:	p-Dichlorobenzene, sodium sulfide
Polymerization:	Polycondensation
Major Uses:	Electrical and mechanical goods
Major Producers:	Bayer, Phillips (Ryton)

Polyphenylene sulfide is a crystalline engineering thermoplastic that can be used with good retention of its physical properties up to its melt temperature, around 300 °C (short time). Increases in modulus and flexural strength can be achieved with glass reinforcement. PPS has good flame-retardant properties and chemical resistance.

The polycondensation of p-dichlorobenzene with sodium sulfide is conducted in a polar solvent. The virgin polymer is relatively low in molecular weight, and it can be used as such in coating applications. Heat curing of PPS in the presence of oxygen affords higher molecular weights as a result of oxidative chain extension and crosslinking. The "cured" PPS is marketed for molding applications.

Major uses for PPS are in the injection molding of electrical (sockets and connectors), electronic (component encapsulation), automotive, and specialty industrial (grilles for hair dryers, microwave oven components, etc.) parts, which require heat or chemical resistance. The coating powders are used in electrostatic coating applications.

The global market for PPS was 16,000 tons in 1991.

17 Heat-Resistant Polymers

Polysulfones

Monomers:	Diphenylsulfonylchloride, 4-chlore-4'-hydroxydiphenylsulfone, bisphenol A, 4,4'-dichlorodiphenylsulfone
Polymerization:	Polycondensation
Major Uses:	Mechanical parts, small appliances, electrical connectors
Major Producers:	Amoco (Udel), BASF (Ultrason), ICI (Victrex), 3M (Astrel)

Polyslufones are a family of engineering plastics with excellent high temperature properties. In 1965 Union Carbide introduced Udel polysulfone in the United States. This business was recently sold to Amoco. Udel has a continuous use temperature of 150 °C and a maximum use temperature of 170 °C, and it can be fabricated easily in conventional equipment. In 1967 the 3M Corporation introduced Ardel, an especially high perfor-

Table 43 Commercial Polysulfones

Trade Name	Polymer Unit	T_g (°C)	Processing
Astrel (3M Corp.)		285	Injection molding in specially modified machines
Polyether sulfone 720 P (ICI)		250	Injection molding with some difficulties in conventional machines
Polyether sulfone 200 P (ICI)		230	Injection molding in conventional machines
Udel (Union Carbide)		190	Injection molding in conventional machines

mance thermoplastic, which requires specialized equipment with extra heating and pressure capabilities for processing. This business was later sold to Carborundum. ICI's polyether sulfones, introduced in 1972, are intermediate in performande and processing. The structures and glass transition temperatures of several commercial polysulfones are listed in Table 43.

The polysulfones are made by either a Friedel-Crafts type polymerization or by nucleophilic polycondensation. For example, Astrel can be made by the Friedel-Crafts reaction of diphenylsulfonylchloride in the presence of aluminum chloride.

Polyether sulfone is obtained by the homopolymerization of 4-chloro-4'-hydroxydiphenylsulfone, which can be obtained by partial hydrolysis of 4,4'-dichlorodiphenylsulfone.

Udel is made by the polycondensation of 4,4'-dichlorodiphenylsulfone with the sodium salt of bisphenol A. This reaction is conducted in highly polar solvents, such as dimethyl sulfoxide or sulfolane.

Polysulfones are among the higher priced engineering thermoplastics. Their use temperature, which is about 160 °C for Udel, is surpassed only by the aramides and polyamide imides, which are not injection moldable, and by the more expensive polyarylketones and liquid crystal polymers. The world market for all engineering plastics in 1990 was about 2 million metric tons. Figure 48 shows the use temperatures of the major engineering thermoplastics. Interestingly, the engineering thermoplastics are generally sold on price/performance basis as shown in Table 44, and performance has only a minor effect on the volume.

17 Heat-Resistant Polymers

Figure 48 Use temperatures of the major engineering thermoplastics

Table 44 Cost and Consumption of Major Engineering Plastics

Engineering Plastic	U.S. Consumption 1989 (tons)	Cost per Cubic Inch (cents)
ABS	600,000	3 – 5
PBT, PET	100,000	5 – 7
PPO	70,000	4 – 7
Polyacetal	60,000	5 – 8
Nylons	230,000	5 – 10
Polycarbonate	200,000	6 – 10
Specialty plastics*	22,000	8 – 18
Blends and alloys	75,000	–

* PPS, polysulfone, PEEK, PEK, liquid crystal polymers

Higher performance engineering thermoplastics, such as polysulfone, PEEK, PEK, and liquid crystal polymers (LCPs) are replacing metals. Since they can be injection molded into complex shapes, they do not require costly machining and finishing operations. They can also be extruded into film and foil, which is of interest for flexible circuitry because of the high temperature performance of these polymers.

Polybenzimidazole (PBI)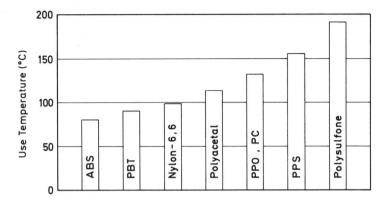

Monomers:	Tetraminobiphenyl, terephthalic acid
Polymerization:	Polycondensation
Major Use:	Fiber
Major Producer:	Hoechst Celanese

Polybenzimidazole has the highest thermal stability of any organic polymer. Its thermal stability (use temperature about 400 °C) combined with good chemical stability indicates that heterocyclic polymers are outstanding candidates for heat-stable polymers. However, their relatively high cost has hitherto prevented commercialization. The synthesis of PBI is carried out as follows:

The tetraminobiphenyl required for the synthesis of polybenzimidazole is obtained from 3,3'-dichloro-4,4'-diaminodiphenyl (a dye intermediate) and ammonia.

The organic polymers with the highest continuous use temperature (about 300 °C) manufactured today are polyimides (see Chapter 15). However, further exploration of heterocyclic polymers could push the use temperature significantly higher, especially if less costly processes are developed for the required monomers.

Liquid Crystal Polymers (LCPs)

Monomers: 4,4'-Dihydroxydiphenyl, p-hydroxybenzoic acid, terephthalic acid, hydroxynaphthoic acid
Polymerization: Polycondensation
Major Uses: Automotive, electronic
Major Producers: Dartco (Xydar), Hoechst Celanese (Vectra)

Liquid crystal polymers are aromatic polyesters. They are self-reinforcing polymers, which have highly ordered structures in the melt and in the solid state. Xydar was introduced in the United States in 1984, and Vectra was launched in 1985. These polymers are the most recent entries into industrial polymers. LCPs are manufactured by the reaction of the acetates of the phenolic monomers with the carboxylic acid group of the same monomer or a comonomer (terephthalic acid).

Xydar is based on 4,4'-dihydroxydiphenyl, p-hydroxybenzoic acid, and terephthalic acid, while Vectra is made from p-hydroxybenzoic acid and hydroxynaphthoic acid. Vectra offers high tensile strength and good notched Izod impact (10 ft-lb/in.). It has low smoke and toxicity, broad chemical inertness, good electrical properties, and ready processability in standard equipment. Xydar has generally slightly lower mechanical properties but a higher heat distortion temperature. LCPs also have outstanding thermal oxidative stability. They are mainly used in demanding electrical, electronic, and automotive applications.

18 Silicones and Other Inorganic Polymers

Introduction

Inorganic polymers could provide temperature and fire resistance vastly superior to those obtainable from organic macromolecules. Numerous inorganic polymers have been synthesized, but only a few have found commercial acceptance because of the difficulties encountered in processing them. Inorganic polymers are mostly insoluble and do not melt under normal processing conditions.

Inorganic polymers are generally classified as chain, sheet, or network polymers. Typical organometallic chain polymers include silicones, polyphosphazenes, polycarborane-siloxanes, and polysiloxanes; these compounds can also exist as cyclic oligomers. A sheet polymer is a macromolecule in which primary valences of the atoms are satisfied by covalent bonds that form a two-dimensional lattice. Typical examples include graphite and boron nitride. Network polymers are macromolecules in which crosslinking results in formation of a three-dimensional network. These polymers are usually hard, infusible, and insoluble substances, such as diamond.

Inorganic homopolymers are polymers in which the repeating unit contains only one element, such as in graphite, where the repeating unit is carbon. However, most inorganic polymers are heteropolymers; that is, the repeating unit contains two or more different elements. A typical example of an inorganic heteropolymer is polyboron nitride. This polymer is obtained readily from borax and ammonium chloride:

$$Na_2B_4O_7 + 2NH_4Cl \xrightarrow{1000\ °C} 2\text{-[B=N]}_n + B_2O_3 + 2NaCl + 4H_2O$$

Polyboron nitride has found industrial applications as refractory material, insulator, and abrasive. Polyboron nitride melts at 3000 °C under pressure, and two structural modifications are known. The hexagonal layer structure, similar to graphite, is used in insulators, and the tetrahedral structure, which is isomorphic and isoelectronic with diamond, is used as an abrasive.

Ceramics, based on clays, silica, and feldspar, are of importance in the electronic industry. Most oxide ceramics are insulators, and most nonoxide ceramics are semiconductors.

Silicones

$$\left[\begin{array}{c}CH_3\\|\\-Si-O\\|\\CH_3\end{array}\begin{array}{c}CH_3\\|\\-Si-O\\|\\CH_3\end{array}\begin{array}{c}CH_3\\|\\-Si-O\\|\\CH_3\end{array}\right]_n$$

Monomers:	Chlorosilanes
Polymerization:	Polycondensation
Major Uses:	Elastomers, sealants, fluids
Major Producers:	Dow Corning, G.E., Union Carbide, Wacker

By far the most important inorganic polymers are based on silicon, an element abundantly available on our planet. The development of an economically attractive process for chlorosilanes, discovered by E. G. Rochow in 1940 at the G.E. Research Laboratories, allows the production of a broad variety of products from a few basic monomers. The basic chemistry involved in Rochow's direct process is as follows:

$$SiO_2 + 2C \longrightarrow Si + 2CO$$

$$Si + 2CH_3Cl \longrightarrow (CH_3)_2SiCl_2 \xrightarrow{H_2O} [(CH_3)_2SiO]_n + 2HCl$$

This is an idealized version of the route from sand (SiO_2) to silicone polymers.

In the alkylation of silicium with methyl chloride, mono-, di-, and trialkyl products are formed. The monoalkyl derivatives produce highly crosslinked gels:

$$RSiCl_3 \xrightarrow{H_2O} \text{highly crosslinked gels}$$

From the dialkyl derivatives linear polymers are obtained:

$$R_2SiCl_2 \xrightarrow{H_2O} -O-Si\left[\begin{array}{c}R\\|\\-O-Si-\\|\\R\end{array}\right]_n$$

The trialkyl derivatives form disiloxanes:

$$R_3SiCl \xrightarrow{H_2O} R_3Si-O-SiR_3$$

The hydrolysis and neutralization of the chlorosilane monomers may be performed by either batch or continuous processes. The continuous process is used for high volume products, such as dimethyl fluids, or room temperature vulcanizing rubbers (RTV).

The silicon products were first developed for high temperature insulation and for encapsulation of electrical parts, but their relative inertness led to the emergence of new applications in the biomedical field. Today silicone products range from liquid fluids to reinforced rigid plastics.

18 Silicones and Other Inorganic Polymers

In the preparation of silicon fluids it is necessary to use highly purified dialkyldichlorosilanes and monofunctional trialkylchlorosilanes as chain stoppers to keep the viscosity of the polymeric liquids constant. These fluids have the advantage of oxidative stability, low freezing point, and a small temperature coefficient of viscosity.

Silicone elastomers are produced from linear polymers based on polydimethylsiloxane by partial crosslinking. The crosslinking mechanisms used include free radical initiated crosslinking, incorporation of trifunctional monomers, and incorporation of unsaturated groups for further reaction. Usually 10 % of the methyl groups in polydimethylsiloxane are replaced by vinyl groups, and such copolymers not only allow crosslinking with peroxide but may also be used as substrates for grafting. It was observed that 3 % silicone rubber equaled 8 % of polybutadiene in toughening, probably due to its low T_g of $-130\,°C$.

Bouncing putty is based on a polydimethylsiloxane polymer modified with boric acid, additives, fillers, and plasticizers to give a material that behaves like an elastic material upon shock but flows like a viscous liquid on slow application of pressure.

Many different silicone products are available today. The major applications are listed in Table 45.

Table 45 Major Applications of Silicones

Hydraulic fluids	Greases and waxes
Heat exchange fluids	Insulation
Antifoaming agents	Dielectric encapsulation
Glass sizing agents	Gaskets and seals
Coupling agents	Caulking agents (RTV)
Surfactants	Rubber molds
Water repellents	Biomedical devices
Masonry additives	Cosmetics

Polysilanes are polymers composed of disubstituted silicon atoms. The relatively low strength of the silicon-silicon bonds results in lower thermal stability. Polysilanes are prepared from calcium silicide and HCl-ethanol in an inert atmosphere. Pyrolysis of polysilanes affords microcrystalline β-silicon carbide (SiC) fibers.

Silicon-nitrogen polymers are obtained from dialkyldiaminosilanes and diamines. The reaction proceeds in two stages.

$$\text{EtNH}-\underset{\underset{R}{|}}{\overset{\overset{R}{|}}{\text{Si}}}-\text{NHEt} + \text{H}_2\text{N}-\text{R}'-\text{NH}_2 \xrightarrow[-2\text{EtNH}_2]{130 \text{ to } 250\,°C} \text{EtNH}-\underset{\underset{R}{|}}{\overset{\overset{R}{|}}{\text{Si}}}-\text{NH}-\text{R}'-\text{NH}-\underset{\underset{R}{|}}{\overset{\overset{R}{|}}{\text{Si}}}-\text{NHEt}$$

$$\xrightarrow[-\text{EtNH}_2]{400\,°C} \text{EtNH}-\underset{\underset{R}{|}}{\overset{\overset{R}{|}}{\text{Si}}}-\text{NH}-\text{R}'-\left[\begin{array}{c} \text{R} \quad \text{R} \\ \text{Si} \\ \text{N} \qquad \text{N}-\text{R}' \\ \text{Si} \\ \text{R} \quad \text{R} \end{array}\right]_n-\text{NH}-\underset{\underset{R}{|}}{\overset{\overset{R}{|}}{\text{Si}}}-\text{NHEt}$$

In the first stage linear polymers are produced, which upon further heating give polycyclodisilazanes.

Polyphosphazenes (PNF)

$$\left[-N=P \begin{array}{c} OCH_2CF_2CF_2H \\ | \\ | \\ OCH_2CF_2CF_2H \end{array} \right]_n$$

Monomers: Phosphorus pentachloride, ammonium chloride, fluorinated alcohols
Polymerization: Polycondensation followed by nucleophilic displacement of chloro groups
Major Uses: Aerospace, military, oil exploration applications
Major Producer: Firestone (PNF)

Nitrogen/phosphorus-containing polyphosphazenes have been synthesized using the inexpensive phosphonitrilic chlorides as starting materials. The general synthetic scheme is as follows:

$$PCl_5 + NH_4Cl \longrightarrow \left[\begin{array}{cc} Cl & Cl \\ | & | \\ -N=P-N=P- \\ | & | \\ Cl & Cl \end{array} \right]_n \xrightarrow{RONa} \left[\begin{array}{cc} OR & OR \\ | & | \\ -N=P-N=P- \\ | & | \\ OR & OR \end{array} \right]_n$$

The chloro groups are subsequently displaced by alkoxy or fluoroalkoxy groups. Using mixtures of alkoxy substituents with longer alkyl chains, crystallization can be avoided, providing amorphous rubbers. Poly(trifluoroethoxy heptafluorobutoxy phosphazene) has a T_g of –77 °C. The rubbers can be further crosslinked with free radical initiators or by radiation.

The polyphosphazene elastomers have excellent resistance to oil, fuels, and chemicals (except alcohols and ketones), good mechanical properties, and a broad temperature service range from –65 to 117 °C. Polyphosphazenes are used in the preparation of O-rings, gaskets, and hydrocarbon fuel hoses.

Polycarborane-siloxanes

$$\left[\begin{array}{c} R \\ | \\ -Si-CB_{10}H_{10}CSi \\ | \\ CH_3 \end{array} \begin{array}{c} R \\ | \\ \\ | \\ CH_3 \end{array} \left(\begin{array}{c} R^1 \\ | \\ Si-O \\ | \\ CH_3 \end{array} \right)_y \right]_x$$

Monomers: Carboranes, dialkyldichlorosilanes
Polymerization: Hydrolytic polycondensation
Major Uses: Liquid phases in gas chromatography, gaskets, wire coatings

Polycarborane-siloxanes are linear polymers based on carboranes. Carboranes ar polyhedral compounds containing boron, hydrogen, and carbon. The carboranes are reacted via their lithium derivatives with dialkyldichlorosilanes, and the resultant oligomers undergo hydrolytic polycondensation to give polycarborane-siloxanes.

The amorphous polycarborane-siloxanes are thermally stable to 400 °C and they have useful elastomeric properties. They can be formulated with fillers and other additives, and they can be vulcanized using standard silicone technology. Their use temperature exceeds 300 °C, and they are perhaps the most thermally stable elastomers prepared so far.

Polythiazyl $\quad -[-S=N-]_n-$

Monomers: S_4N_4
Polymerization: Ring-opening polymerization
Major Use: Electrode material

Poly(sulfur nitride) or polythiazyl is obtained by pyrolysis of gaseous S_4N_4 under vacuum. The solid four-membered ring reaction product S_2N_2 is consecutively polymerized at room temperature by ring-opening polymerization to give the linear chain polythiazyl, a crystalline, fibrous material, which is soft and malleable. The synthesis of polythiazyl can be represented as follows:

$$S_4N_4 \xrightarrow[\text{1 torr}]{300\,°C/Ag} \begin{array}{c} S-N \\ | \quad | \\ N-S \end{array} \xrightarrow{\text{room temperature}} -[-S=N-]_n-$$

Polythiazyl has electrical conductivity at room temperature and superconductivity at 0.3 K. Polymeric electroconductors are preferred over metals because they are flexible and lighter in weight. Upon doping, polythiazyl can become a photoconductor (see Chapter 19).

19 Functional Polymers

Introduction

Today's major polymers exist because of physical properties and economics. The large volume polyolefins are rather dull macromolecules that can best be described as polyparaffins. In contrast, polypeptides, the building blocks of living matter, are highly sophisticated functional macromolecules. In the closing years of this century much effort in polymer science will focus on functional polymers, whose value lies in their reactivity rather than their physical properties. For example, simple functional polymers are ion-exchange resins and hollow-fiber polymer membranes. The U.S. troops during the Gulf War were supplied with drinking water made by desalination of seawater, using hollow-fiber polymer membranes. More sophisticated examples include photopolymers, electroconductive polymers, optical fibers, and biopolymers made by genetic engineering. Perhaps the most dramatic example of this trend is electrophotography, which is often referred to as xerography. This technology has revolutionized human communication in the last 50 years.

Photoconductive Polymers

Electrophotography was invented by Chester F. Carlson in 1937, but the first commercial units were sold as late as 1950, because the significance of this new technology was not recognized by major companies in the information business. Today xerography is everywhere. The steps in the electrophotographic process are as follows.

- Sensitization of a photoconductive surface (selenium or polyvinylcarbazole) to light by electrostatic charging in darkness
- Imagewise exposure
- Development of the latent image with toner
- Transfer of image to paper
- Fixing of the image on the paper with heat
- Removing residual powder from the drum

Electrophotography uses selenium as the imaging material, but organic polymers, such as polyvinylcarbazole with added chemical (Lewis acids) and optical (dye molecules) sensitizers compete effectively today.

Electroconductive Polymers

Organic solids are regarded as poor electrical conductors, usually behaving as electric insulators. However, new classes of organic solid conductors (donor-acceptor complexes) as well as some organic and inorganic polymers have emerged with electrical conductivities approaching that of typical metals.

In 1977 Shirakawa in Japan and A. G. MacDiarmid and A. J. Heeger in the United States discovered that partial oxidation of polyacetylene films with iodine increased their photoconductivity significantly. The transformation of a polymer to its conductive form via chemical oxidation or reduction is called doping. Polyacetylene can exist in trans or cis configurations, and the trans structure has the highest conductivity. All electroconductive polymers contain conjugated double bonds and typical examples are listed in Table 46. Processing of polyacetylene, polypyrrole, polythiophene, and the like, is not possible because these polymers do not melt or dissolve. 3-Alkyl substitution in polythiophene increases its solubility and fusibility. The amorphous electroconductive polymers made so far contain imperfections. Efforts are under way to synthesize polymers with fewer imperfections and increased crystallinity in order to increase their electroconductivity. Currently conductivities can be as high as 10,000 S/cm in the case of doped and stretch-oriented polymers. The conductivity of copper is 6×10^8 (S/cm).

Table 46 Semiconductive Polymers and Charge Transfer Complexes

Name	Structure	Conductivity, σ_{RT} (Siemens/cm)
Polyacetylene		10,000*
Poly(3-alkylthiophene)		10,000*
Polyphenylene vinylene		10,000*
Polypyrrole		7,500
Polyphenylene**		5,000

(continued on next page)

Name	Structure	Conductivity, σ_{RT} (Siemens/cm)
Polythiazyl	$-[S=N]_n-$	3,700
Polythienyl vinylene	(thiophene-vinylene)$_n$	2,700*
Polyphenylene sulfide	$-[C_6H_4-S]_n-$	500
TCNQ*** salt of TTF****	TCNQ · TTF	500
Polyaniline	$-[C_6H_4-NH]_n-$	200*
Polyisothianaphthene	(isothianaphthene)$_n$	50

* As oriented polymer.
** Made and doped in situ by treating a layer of crystalline *p*-terphenyl on a glass plate with arsenic pentafluoride.
*** 7,7,8,8-Tetracyanoquinodimethane (acceptor).
**** Tetrathiafulvalene (donor).

Electroconductive polymers are used in the construction of plastic thin film batteries. For example, two sheets of stereoregular polyacetylene separated by an insulating membrane, soaked in lithium perchlorate and using propylene carbonate as electrolyte, can be laminated between plastic sheets. The anode-charging reaction converts one polyacetylene sheet to a polyanion with lithium counterions. The cathode charges by conversion of the other sheet to a polycation with perchlorate counterions. Such a battery delivers 3.7 V and 5 mA/cm^2. Other polymers used for longlife batteries are polyaniline, polypyrrole, or polythiophene as the cathode. The anodes are usually lithium or lithium alloys (lithium-aluminum). Solid polymer electrolytes, such as lithium salt/polyethylene oxide intended for an all-solid-state battery are under development. Other potential uses include incineration-disposable electrostatic precipitator screens; heating elements for wall coverings, floors, blankets, or clothing; and noncorrosive electrodes.

Polyaniline shows a range of colors as a result of protonation or oxidation. The electrochromic properties of polyaniline, which is yellow in its undoped form, but green or blue in its doped forms, are used in displays and thermal "smart windows" (which darken during summer to absorb some of the sunlight, thereby saving air conditioning costs). In a smart window, both electrodes are transparent. To block sunlight, a positive potential is applied, causing oxidative doping of the polymer. To reverse the process, a negative potential is applied.

Piezoelectric Polymers

The most effective piezoelectric (generation of electricity on deformation) polymer is polyvinylidene fluoride. The polymorphism of polyvinylidene fluoride and the two distinct dipole groups (CF_2) and (CH_2) contribute to the polymer's exceptional dielectric properties. Polyvinylidene fluoride is used widely in microphones, earphones, loudspeakers, burglar alarms, and fire detection devices. Nylon-11 is another commercially available polymer with piezoelectricity about half that of polyvinylidene fluoride and the same pyroelectricity. The electrical properties of nylon-11 are explained by noting that nylons with odd numbers of carbons have net dipole moments per unit cell. Orientation of the dipoles in films of semicrystalline polymers by subjecting them to electrostatic fields of 5000 kV/cm produces materials that can be used in infrared-sensitive television cameras and in underwater detection devices for submarine. Polymer films can be used easily for fabrication into electronic devices because they can be overlaid with printed circuits.

Light-Sensitive Polymers

While electrophotography produces a temporary image, permanent images can be produced be photopolymerization or photocrosslinking. In photopolymerization processes, monomers undergo selective polymerization upon exposure to light. Typical examples include Du Pont's Lydel and Dycryl systems. In photocrosslinking processes, soluble linear polymers become crosslinked and insoluble upon exposure to ultraviolet light.

A typical example of photocrosslinkable polymers is offered by the polyvinyl cinnamates marketed by Eastman Kodak. Polyvinyl cinnamates are manufactured by esterification of polyvinyl alcohol with cinnamic acid. The photoreactive groups can be pendant to the polymer chain as in polyvinyl cinnamates, but there are also systems in which the photoreactive group is part of the polymer chain. A third approach involves the addition of difunctional photoreactive additives to a base polymer. The types of crosslinking mechanism operating in photopolymers are depicted in Figure 49.

Pendant Photoreactive Double Bonds

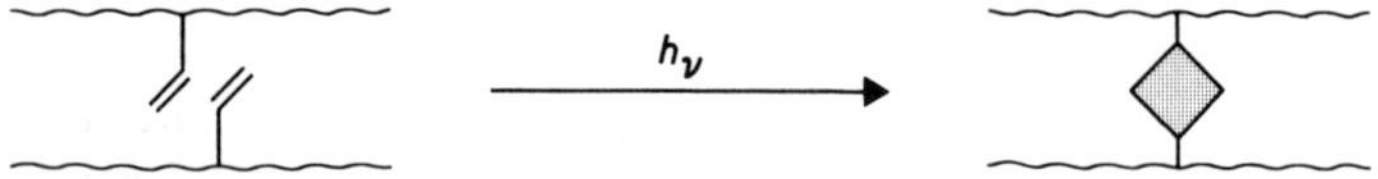

Photoreactive Double Bonds as Part of Chain

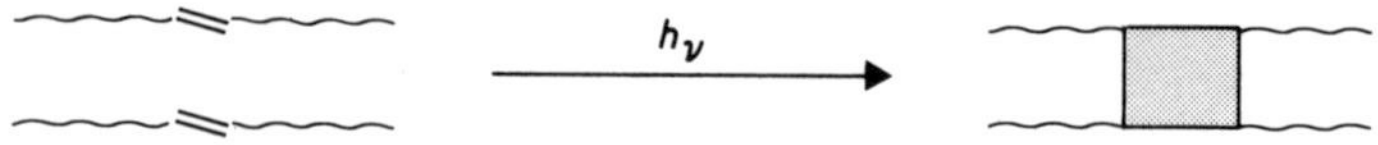

Difunctional Photocrosslinker

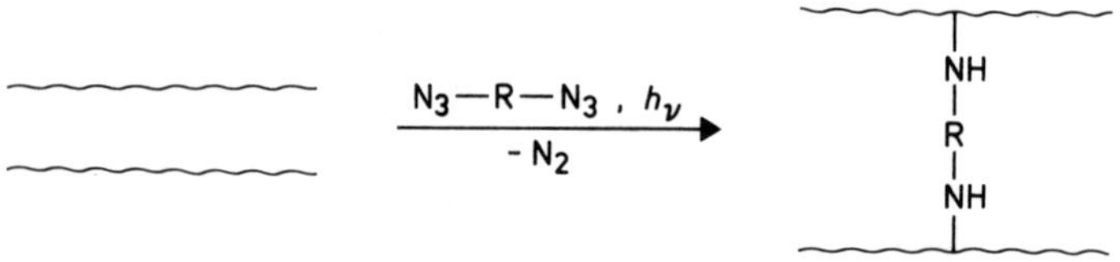

Figure 49 Photocrosslinking mechanisms

Difunctional azides (see Figure 49) could also be used to generate bubble (vesicular) images in a polymer matrix.

The photopolymers are used in photofabrication (photoresist, photomilling) and for the formation of printing plates. Microimaging, using photopolymer technology, has revolutionized the electronic industry. For example, microcircuits for electronic applications are easily mass-produced by first applying the linear polymer, followed by imagewise exposure, which leads to crosslinking and insolubilization of the exposed area. Solvent removal of the unexposed material and etching generates the microcircuit. Plastic printing plates are produced in a similar manner, thereby eliminating the classic typesetting process that involves the melting of lead. Most newspapers today are printed from photopolymer printing plates.

Photocrosslinkable systems are negative working because the exposed area is crosslinked. Exposed areas can also be selectively solubilized using positive working systems. For example, addition of quinonediazides to a polymer matrix leads to the generation of basesoluble carboxyl groups:

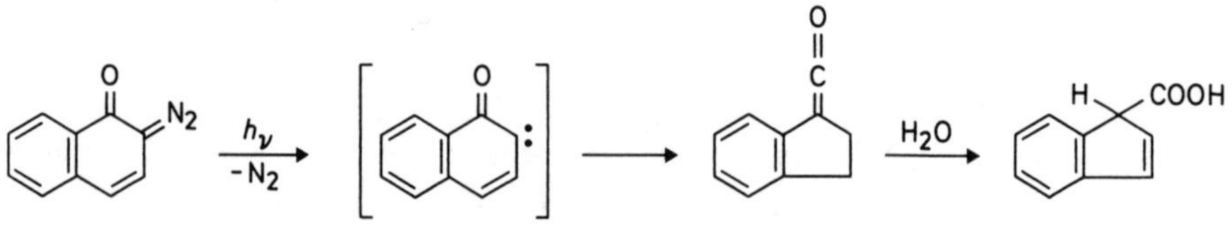

Further advances in technology could lead to non-silver-halide photography. The resolution of photopolymer systems is far superior to that obtainable from silver halide, because the limit in resolution is provided by the size of the macromolecules. In contrast, the silver grain produced in the development of a silver halide film is considerably larger. For example, the total content of the Bible can be reproduced on a standard 35 mm negative using photopolymers.

Hollow-Fiber Membranes

Hollow-fiber membranes are used in reverse osmosis systems for the desalination of brackish water and seawater. Other applications include ultrafiltration/microfiltration, hemodialysis (artificial kidneys), and systems for gas separation, including the recovery of hydrogen in ammonia plants and in oil refineries, the production of nitrogen from air, and the separation of carbon dioxide and hydrogen sulfide from natural gas. Another important new membrane technology is pervaporation, which allows separation of liquid organics from other liquid organics. An example is the dehydration of ethanol, which is important in gasohol production.

A hollow-fiber membrane is a capillary having a diameter of less than 1 mm and whose walls function as a semipermeable membrane. Usually hollow-fiber materials are used as cyclindrical mebranes, which offer enhanced productivity per unit volume over flat sheet or tubular membranes. The polymers used in the construction of hollow-fiber membranes are cellulose acetate, polysulfones, polyacrylonitrile, polymethyl methacrylate, polyamides, and polybenzimidazole.

During the Gulf War, the U.S. Army's mobile purification units, which supplied all the drinking water for the troops, used reverse osmosis technology. Treatment of municipal wastes, agricultural runoff, textile wastes, and wastes from chemical operations also can be conducted using this technology.

Ion-Exchange Resins

The most extensively used functional polymes are ion-exchange resins. Modern ion-exchange technology began between 1935 and 1940 with the pioneering work of Adams and Holmes in England. Today more than 100 synthetic ion-exchange resins are marketed throughout the world. They are broadly classified as cation- and anion-exchange materials; their basic chemical structures are shown in Table 47.

Table 47   Types of Ion-Exchange Resin

Active Group	Structure
Cation-Exchange Resins	
Sulfonic acid	⌬—SO₃H

Carboxylic acid	—CH₂CHCH₂—
	          │
	       COOH

Phosphonic acid	⌬—PO(OH)₂
Anion-Exchange Resins	
Quaternary ammonium salt	⌬—CH₂N(CH₃)₃]⊕ Cl⊖
Secondary amine	⌬—CH₂NHR
Tertiary amine	⌬—CH₂NR₂

Sulfonated copolymers of styrene and divinylbenzene constitute the most widely produced type of cation-exchange resin. Typical products are Amberlite IR-120, Dowex 50, Duolite C-20, and Ionac C-240. The sulfonation of the base polymer is conducted in a batch process, and it is attempted to introduce one sulfonic acid group into the para position of each aromatic nucleus. Weak acid, cation-exchange resins are based on acrylic and methacrylic acid. Introduction of phosphonic acid groups is best accomplished by reacting polystyrene with phosphorus trichloride in the presence of aluminum chloride.

The anion-exchange resins are obtained via chloromethylation of polystyrene, followed by amination. Also phenol-formaldehyde resins are used as base polymers for the manufacture of ion-exchange resins.

Applications of ion-exchange polymers are extremely varied, ranging from water treatment to the purification of chemicals, such as sugar and formaldehyde. Ion-exchange resins are also exceedingly useful in the analysis of complex mixtures of products. Even wines are sometimes treated by column cation exchange to prevent precipitation of potassium bitartrate during storage. Finely divided ion-exchange resins have been used as therapeutic agents to reduce gastric acidity or to neutralize bile acids.

A new ion-exchange membrane for chloralkali production was designed by copolymerization of tetrafluoroethylene with a vinyl ether perfluoroacid precursor (Figure 50). Subsequent hydrolysis produced the membrane material, which has high conductivity and permeability for sodium ions and excellent chemical resistance toward chlorine and strong caustic.

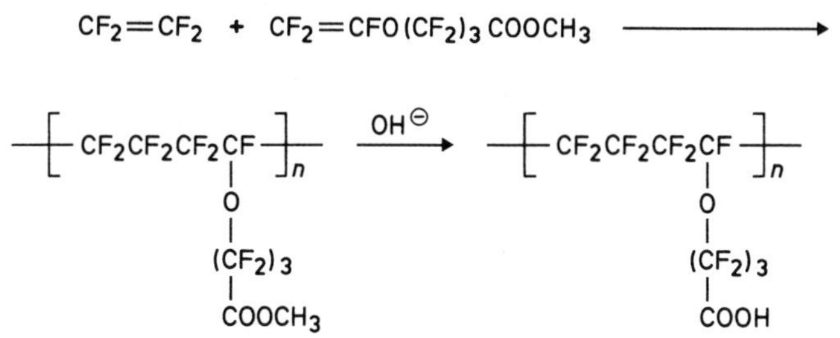

$$CF_2{=}CF_2 \;+\; CF_2{=}CFO(CF_2)_3COOCH_3 \longrightarrow$$

$$\mathrm{-\!\!\left[CF_2CF_2CF_2CF\right]_{\!\!n}\!\!-} \quad \xrightarrow{OH^{\ominus}} \quad \mathrm{-\!\!\left[CF_2CF_2CF_2CF\right]_{\!\!n}\!\!-}$$
$$\underset{\underset{COOCH_3}{|}}{\underset{(CF_2)_3}{|}}{O} \underset{\underset{COOH}{|}}{\underset{(CF_2)_3}{|}}{O}$$

Figure 50   Polyperfluorocarboxylic acid

Polymeric Reagents

Considerable efforts in research are under way to immobilize catalysts or reagents attached to polymers. Polystyrene with varying degrees of crosslinking is commercially available with pendant chloromethyl or diphenylphosphine groups. Further reaction of such polymers produces reagents with various degrees of swelling in organic solvents depending on the degree of crosslinking. Amino acid esters and steroids can be attached to these polymers for slow release under biological conditions. Controlled release of drugs can also be achieved by impregnating insoluble polymers with therapeutically active ingredients that are slowly extracted by the body fluids, or by encapsulating therapeutic agents with soluble or biodegradable polymers. An example is the use in contraception, where an intrauterine progesteron device made from an ethylene-vinyl acetate copolymer provides controlled release for over one year. Also, polymers can be implanted next to a tumor site. For example, bischloronitrosourea in a polymer matrix can be implanted in the brain next to a tumor site. This new method can cure brain tumors in patients who otherwise would have a zero chance of survival.

Enzymes are immobilized by covalent bonding onto polymer surfaces or by entrapment within a gel or a microcapsule. For example, a solid polyamine is available under the trade name Colestid to be taken with orange juice. This polymer is biologically inert, but the polyamine serves to neutralize and remove bile acids. Removal of bile acids causes the human body to produce more bile acids from cholesterol, thereby reducing the cholesterol level in the bloodstream.

The advantages of drugs attached to polymer backbones are shown in Table 48.

Table 48   Advantages of Polymer Drugs

Delayed action	Alteration and modification of drug and activity by altering the solubility of the carrier
Sustained release	
Ease of targeting the drug	Potentiation of activity of drug use in combination with polymer*
Potential of coupling one or several drugs to the same polymer	Reduction of side effects

* For example, potentiation of tetracyclines by complexing with polyacrylic acid. The polymeric acid serves to promote greater absorption of the antibiotic into the bloodstream

The use of polymers as materials of construction for biomedical devices is a growing and important new area of application for biologically compatible polymers. Perhaps the most exciting example of such an application is the polyurethane heart used as a temporary device in heart transplant surgery. Research is under way to further improve this concept to create polymeric hearts that can be used for longer periods of time. Polyurethanes are also used in heart assist devices (pacemakers) and in the construction of artificial blood vessels.

Polymer-bound catalysts and reagents are used extensively today. The workhorses for constructing polymeric reagents are the so-called Merrifield resins (chloromethylated polystyrene), in which the reactive chloromethyl group can undergo nucleophilic displacement reactions with a wide variety of nucleophiles. Polymeric catalysts have the advantage of allowing easy removal from the reaction mixture. Previously, homogeneous metal catalysts had to be reclaimed from the reaction residues. In the construction of polymeric hydrogenation catalysts, selective hydrogenation of small olefins can be achieved because larger molecules are not able to penetrate the polymer to reach the active catalyst sites.

Resin-bound chemical reagents also offer ease of separation and a utility similar to the ion-exchange resins (see Ion-Exchange Resins, above). For example, polymer-bound carbodiimide can be used for the synthesis of anhydrides from carboxylic acids and aldehydes from labile alcohols. The carbodiimide function is converted to the corresponding urea, which is readily separated and reconverted back into polymer-bound carbodiimide:

$$\text{(P)}-\text{C}_6\text{H}_4-\text{CH}_2\text{N}=\text{C}=\text{NR} \xrightarrow{\begin{array}{c}\text{R'COOH}\\\text{R''CH}_2\text{OH}\end{array}} \begin{array}{c} \text{R'COOCOR'} + \text{(P)}-\text{C}_6\text{H}_4-\text{CH}_2\text{NHCONHR} \\ \text{R''CHO} + \text{(P)}-\text{C}_6\text{H}_4-\text{CH}_2\text{NHCONHR} \end{array}$$

Polymeric Wittig reagents are especially useful when an excess of the analogous monomeric reagent, or its reaction product, causes separation problems. Sterically hindered resin-bound sulfonyl chlorides were used in the synthesis of oligonucleotides.

Also, low effective reagent concentration can be achieved using polymeric reagents. Numerous other applications are known, and the use of macromolecules in effecting selective chemical reactions could bring us closer to the perfection achieved in biological systems.

The synthesis of starburst dendrimers is an example of building synthetic polymers selectively to achieve the desired end-use applications. Starburst dendrimers contain three or more highly branched polymer chains, which orginate from a tiny core:

Hyperbranched polymers were first synthesized in the late 1970s, but it was not until the mid-1980s that D. A. Tomalia at Dow Chemical found a way of making them with a uniform molecular weight and size. The first dendrimers were polyamidoamines (PAMAM) grown from a threebranched core prepared by reacting ammonia with methyl acrylate, followed by an excess of ethylenediamine. At the end of each branch is a free amino group, which can react with two more methyl acrylate monomers followed by excess ethylenediamine. These steps were repeated to form successive generations. The ninth-generation PAMAM contains 3069 monomer units and has a molecular weight of 349,883 and a diameter of 123 Å. The number of end groups is 1536. Many different dendrimer families with different properties were made. The porosity of the center membrane can be controlled. Some dendrimers are tightly sealed and other are fairly leaky.

Dendrimers can be used in microelectronics, in the calibration of sieves, and as contrast agents for NMR imaging. For the latter application, gadolinium or manganese ions were chelated on the dendrimer surfaces. Dendrimers also have an enormous surface area in relation to volume. The combination of water solubility and high surface area could make these polymers useful as catalyst carriers.

Biopolymers

The three major classes of natural macromolecules are proteins, nucleic acids, and polysaccharides. They are formed by stepwise condensation of monomer units, with formation of water as the by-product. The sequence of the monomeric units along the chain represents the primary structure of the biopolymer. Secondary structures result from the interaction of monomeric units with each other along the chain. An example is the helical structure of DNA caused by hydrogen bonding between groups along the same chain. Also sheetlike arrays between groups on adjacent chains occur. Chains can also form tertiary structures in which the chains are folded into complex three-dimensional forms.

Proteins, made from amino acids, are the most abundant of natural biopolymers, and nucleic acids control the growth of living systems. Polynucleotides are made from repeating sequences of nucleotides joined by phosphodiester linkages. The polynucleotides are subdivided into two classes depending upon the nature of their sugar residues. Ribonucleic acids (RNAs) contain β-D-ribose, and deoxyribonucleic acids (DNAs) contain β-2-deoxy-D-ribose. DNAs store the genetic information in living systems, and RNAs assist in the transfer of the genetic information. Polysaccharides are important for their structural and fuctional roles in the cell walls of bacteria and plants, in the cell membranes of animals, and also as structural components in plants. Starch, a polysaccharide, is the most common fuel storage material in plants.

Synthetic analogues of proteins and nucleic acids are synthesized to elucidate secondary structures in biopolymers. Protein (peptide) synthesis uses solid phase stepwise polymerization techniques conducted on polymer resins. Examples are slightly crosslinked functional polystyrene resins containing 0.5 to 1.0 % divinylbenzene (Merrifield resins; see also polymeric reagents). The chloromethylated resins can directly react with the salts of amino acids blocked on the nitrogen function to afford an ester bond to the polymeric substrate. Unblocking of the amino function and reaction with another nitrogen-blocked amino acid gives the first amide bond. This step can be repeated many times, and commercial peptide synthesizers can be used to synthesize polypeptides of up to 120 amino acid units. The biopolymer is removed from the solid phase using HF as the reagent.

Reaction of the chloromethyl group in the Merrifield resins with ethylene glycols affords functional hydroxyl groups attached to the polymeric substrates. Nitrogen-blocked amino acids can now be attached by esterification using dicyclohexylcarbodiimide as the reagent. Also ethylene oxide can be grafted to the hydroxyl-terminated polymer to give crosslinked polystyrene containing polyethylene glycol units. Another approach to the construction of a reactive solid substrate is the reaction of crosslinked polystyrene with *p*-nitrobenzoyl chloride in the presence of aluminum chloride, forming polymeric ketones. Subsequent reaction of the keto groups with hydroxylamine gives the corresponding oximes, to which amino acids can be attached in a stepwise manner.

The final purification of the biopolymers is conducted by liquid high-pressure chromatography or by electrophoresis on polyacrylamide gels. Oligonucleotides can also be synthesized using commercial solid phase synthesizers, but polysaccharids cannot as yet be synthesized using such procedures.

Another approach to the synthesis of biopolymers is genetic engineering using bacterial cells or yeasts as substrates. In the first step of such a process cyclic DNA molecules (plasmids) isolated from *Escherichia coli* are treated with a restriction enzyme to selectively remove a segment from the plasmid. The resultant linear DNA is subsequently treated with "passenger" DNA to reclose the ring using a different enzyme (DNA ligase). The passenger DNA can now be amplified by replication of the hybrid molecule in the bacteria. The incorporation of the recombined DNA into the bacteria cells is aided by treatment of the bacterial cells with $CaCl_2$ solution to render the cell walls more permeable. The isolation of the recombinant DNA is the biggest problem, because only one of a thousand cells incorporates the recombinant DNA, and again only a small portion of these cells have the correct conformation.

Recombinant DNA technology is used extensively to produce macromolecular peptides used as biopharmaceuticals. Proteins (polypeptides), such as insulin, human growth hormone, blood factors (proteins involved in the clotting of blood), growth factors (proteins responsible for directing the differentiation and production of various cell types), and erythropoietin (a protein produced in the kidney that stimulates red blood cell production) are already produced as biopharmaceuticals. Also monoclonal antibodies (antibodies that are specific to certain antigens or cell types), interferons (broad-acting agents by a variety of cells), and interleukins (immune system hormones) are on the market.

Appendix

Guide to Further Reading

Part I: General Introduction

Staudinger, H.: From Organic Chemicals to Macromolecules. Wiley Interscience, New York, 1961.
Flory, P. J.: Principles of Polymer Chemistry. Cornell University Press, New York, 1953.
Billmeyer, F. W. Jr.: Textbook of Polymer Science, 3rd ed. Wiley Interscience, New York, 1984.
Elias, H. G.: Macromolecules. Plenum Publishing Corp., New York, 1977.
Collins, E. A., Bares, J., and *Billmeyer, F. W. Jr.*: Experiments in Polymer Science. Wiley Interscience, New York, 1973.
Platzer, N. A. J.: Addition and Condensation Polymerization Processes. In: Advances in Chemistry, Vol. 91. American Chemical Society, Washington, D. C., 1969.
Sandler, S. R., and *Karo, W.*: Polymer Synthesis. In: Organic Chemistry, Vol. 29, I-III. Academic Press, New York, 1974, 1977, 1980.
Ward, F. M.: Mechanical Properties of Solid Polymers. Wiley Interscience, New York, 1971.
Ulrich, H.: Raw Materials for Industrial Polymers. Hanser Publishers, Munich, Vienna, New York, 1988.
Tadmor, Z., and *Gogos, C. G.*: Principles of Polymer Processing. John Wiley & Sons, Inc., New York, 1979.
Throne, J. L.: Plastics Process Engineering. Marcel Dekker, Inc., New York, 1979.
Modern Plastics Encyclopedia 1990. McGraw-Hill, Inc., New York, 1990.

Part II: Addition Polymers

Boenig, H. Y.: Polyolefins: Structure and Properties. Elsevier, New York, 1966.
Natta, G., and *Danusso, F.*: Stereoregular Polymers and Stereospecific Polymerizations. Pergamon Press, New York, 1967.
Boor, J. Jr.: Ziegler-Natta Catalyst and Polymerization. Academic Press, Inc., New York, 1979.
Galanti, A. V., and *Mantell, C. L.*: Polypropylene Fibers and Films. Plenum Press, New York, 1965.
Lenz, R. W., and *Ciardelli, F. (Eds.)*: Preparation and Properties of Stereoregular Polymers. D. Reidel Publishers, Dordrecht (NL), 1980.
Cotter, R. J., and *Matzner, M.*: Ring-Forming Polymerization. In: Organic Chemistry, Vol. 13-B, I, II. Academic Press, New York, 1972.
Plesch, P. H.: Cationic Polymerization. Macmillan, New York, 1963.

Reich, L., and *Schindler, A.*: Polymerization by Organometallic Compounds. In: Polymer Reviews, Vol. 12. Interscience, New York, 1966.
Burgess, R. N.: Manufacture and Processing of PVC. Macmillan Publishing C., Inc., New York, 1982.
Eirich, F. R. (Ed.): Science and Technology of Rubber. Academic Press, Inc., New York, 1978.
Noshay, A., and *McGraph, J. E.*: Block Copolymers: Overview and Critical Survey. Academic Press, Inc., New York, 1976.
Paul, D. R., and *Newman, S. (Eds.)*: Polymer Blends, Vols. I and II. Academic Press, Inc., New York, 1978.
Horn, M. B.: Acrylic Resins. Reinhold Publ. Corp., New York, 1960.
Ham, G. E. (Ed.): Vinyl Polymerization, Vol. 1. Marcel Dekker, Inc., New York, 1967.
Gum, W. F., Riese, W., and *Ulrich, H. (Eds.)*: Reaction Polymers. Hanser Publishers, Munich, Vienna, New York, 1992.
Oertel, G. (Ed.): Polyurethane Handbook. Hanser Publishers, Munich, Vienna, New York, 1985.
May, C.: Epoxy Resins. Marcel Dekker, New York, 1988.
Baily, F. E., Jr., and *Koleske, F. V.*: Poly(ethylene oxide). Academic Press, Inc., New York, 1976.

Part III: Condensation Polymers

Foy, G. F.: Engineering Plastics and their Commercial Development. In: Advances in Chemistry, Vol. 96. American Chemical Society, Washington, D. C., 1969.
Meyer, R.: Polyester Molding Compounds and Molding Technology. Chapman & Hall, New York, 1987.
Ludewig, H.: Polyester Fibers, Chemistry and Technology. Wiley Interscience, New York, 1971.
Schnell, H.: Chemistry and Physics of Polycarbonates. Wiley Interscience, New York, 1964.
Kohan, M. I.: Nylon Plastics. Wiley & Sons, Inc., New York, 1973.
Mark, H. F., Atlas, S. M., and *Cernia, E. (Eds.)*: Man-Made Fibers. Interscience Publishers, New York, 1968.
Knop, A., and *Pilato, L.*: Phenolic Resins. Springer Verlag, New York, 1985.
Vale, C. P., and *Taylor, W. G. K.*: Aminoplastics. Iliffe Books, Ltd., London, 1964.

Part IV: Special Polymers

Stone, F. G. A., and *Graham, W. A. G.*: Inorganic Polymers. Academic Press, Inc., New York, 1962.
Cassidy, P. E.: Thermally Stable Polymers. Marcel Dekker, Inc., New York, 1980.

Noll, W.: Chemistry and Technology of Silicones. Academic Press, Inc., New York, 1968.
Kunin, R.: Ion Exchange Resins. R. E. Krieger Publ., Co., Huntington, New York, 1972.
Moyneux, P.: Water Soluble Synthetic Polymers: Properties and Behavior, Vols. I-II. CRC Press, Inc., Boca Raton, FL, 1983.
Seymour, R. B. (Ed.): Conductive Polymers. Plenum Press, New York, 1981.
Roffey, C. G.: Photopolymerization of Surface Coatings. John Wiley & Sons, New York, 1982.
Szycher, M.: Biocompatible Polymers, Metals and Composites. Technomic Publishers, Lancaster, PA, 1983.

Major World Producers of High Volume Industrial Polymers

Polymer	Geographic Location	Company	1990 Capacity (1 000 metric tons)
LDPE	United States	Quantum Chemical	785
		Union Carbide	675
		Dow Chemical	462
		Chevron	385
		Du Pont	320
		Exxon	299
	Western Europe	Neste	800
		Rheinische Olefinwerke	740
		Essochem	480
		Montedison	460
		DSM	430
		Ato Chimie	370
		Dow Chemical	340
	Japan	Sumitomo	230
		Mitsubishi Petrochemical	205
HDPE	United States	Quantum Chemical	800
		Phillips	700
		Soltex	644
		Exxon	590
		Occidental Chem.	580
		Chevron	270
		Dow Chemical	261
	Western Europe	Hoechst	400
		BP	242
		Rheinische Olefinwerke	210
		Montedison	160
		DSM	150
	Japan	Mitsui Petrochemical	200
		Showa Yuka	154
PVC	United States	Shintech	1090
		Occidental Chem.	945
		B.F. Goodrich	788
		Formosa	475
		Tenneco	338
		Georgia Pacific	318
	Western Europe	Montedison	550
		Huels	410
		Rhône-Poulenc	390
		ICI	370
		BASF	280

Polymer	Geographic Location	Company	1990 Capacity (1 000 metric tons)
	Japan	Shin-Etsu	218
		Kanegafuchi	216
		Nippon Zeon	192
PP	United States	Himont	990
		Amoco	760
		Exxon	486
		Fina	405
		Aristech	322
		Shell	270
	Western Europe	Hoechst	520
		Montedison	345
		ICI	245
	Japan	Mitsubishi	190
		Chisso	155
		Mitsui Toatsu	155
PS	United States	Huntsman	578
		Dow Chemical	573
		Arco	545
		Polysar	362
		Fina	288
	Western Europe	BASF	843
		Atochem	455
		Dow Chemical	360
		Montedison	305
		BP	190
	Japan	Asahi-Dow	267
		Idemitsu	200
		Denka	160
		Mitsubishi-Monsanto	115
Polybutadiene	United States	Goodyear	136
		Bridgestone	110
		Phillips	68
	Western Europe	Michelin	72
		Huels	70
		Anic	60
		Shell	50
Polyisoprene	United States	Goodyear	60
	Westen Europe	Shell	75
		Compagnie du Polyisoprene[a]	40
		Anic	30

[a] Joint enterprise, Goodyear/Michelin.

Major World Producers of High Volume Industrial Polymers

Polymer	Geographic Location	Company	1990 Capacity (1 000 metric tons)
	Japan	Japan Polyisoprene[b]	30
		Kuraray	30
		Nippon Zeon	30
ABS	United States	General Electric	450
		Monsanto	325
		Dow Chemical	177
	Western Europe	Bayer	75
		General Electric	70
		BASF	60
	Japan	Japan Synthetic Rubber	60
		Kanegafuchi	48
		Ube Cycon	48
SBR	United States	Bridgestone	405
		Goodyear	342
		B.F. Goodrich	152
		General Tire	138
	Western Europe	Shell	220
		Bayer	202
		Huels	180
Styrene Block Copolymers	United States	Shell	91
		Phillips	34
	Western Europe	Rheinische Olefinwerke	30
		Shell	30
	Japan	Japan Elastomer Company[c]	30
EPDM	United States	Du Pont	77
		Exxon	60
		Uniroyal	44
	Western Europe	DSM	50
		Montedison	50
		Huels	25
Acrylic Fibers	United States	Du Pont	144
		Monsanto	143
		Cyanamid	56
Acrylic Resins	United States	Rohm & Haas	150
		Cyro[d]	36
		Du Pont	35
	Western Europe	Rohm	50
		ICI	32
		Degussa	19
		Montedison	18

[b] Joint enterprise, Japan Synthetic Rubber/Bridgestone. [c] Joint enterprise, Asahi/Showa Denko.
[d] Joint enterprise, Cyanamide/Rohm GmbH.

Major World Producers of High Volume Industrial Polymers

Polymer	Geographic Location	Company	1990 Capacity (1 000 metric tons)
	Japan	Mitsubishi Rayon	56
		Asahi	23
		Mitsubishi Acetate	23
		Kyowa Gas	19
Polyacetal	United States	Hoechst-Celanese	57
		Du Pont	33
	Western Europe	Du Pont	40
		Hoechst	30
	Japan	Polyplastic[e]	35
		Asahi	10
Epoxy Resin	United States	Dow Chemical	77
		Shell	45
		Hoechst-Celanese	30
		Union Carbide	27
	Western Europe	Dow Chemical	35
		Ciba-Geigy	27
		Shell	22
	Japan	Mitsubishi Petrochemical	20
		Asahi-Ciba	16
		Mitsui Kanebo	5
Polyester Fiber	United States	Fiber Ind.[f]	525
		Eastman Kodak	236
		American Hoechst	184
		Akzona	63
Polyester (Unsaturated)	United States	Dainippon Ink (Reichold)	159
		Aristech	136
		Ashland	64
	Western Europe	BASF	82
		Bayer	55
		BP	55
	Japan	Dainippon Ink	60
		Nippon Shokubai	40
		Showa Highpolymer	38
		Takeda	34
PET Film	United States	Du Pont	70
		Eastman Kodak	35
		ICI	34
		3M	32
	Western Europe	ICI	36
		Du Pont	12
		Hoechst	12

[e] Joint enterprise, Daicel/Celanese.
[f] Joint enterprise, Celanese/ICI.

Major World Producers of High Volume Industrial Polymers

Polymer	Geographic Location	Company	1990 Capacity (1 000 metric tons)
	Japan	Toray	26
		Teijin	20
		Fuji	18
PET/PBT Resin	United States	Eastman Kodak	91
		Goodyear	91
		American Hoechst	68
		General Electric	60
	Western Europe	Akzo	25
	Japan	Toray	8.5
PC	United States	General Electric	120
		Mobay	120
		Dow	50
	Western Europe	Bayer	150
		General Electric	60
		Dow Europe	36
		Enichem	20
	Japan	Teijin	15
		Mitsubishi Gas	12
Polyamide Fiber	United States	Du Pont	500
		Monsanto	224
		Allied	180
		Akzona	108
Polyamide Resin	United States	Du Pont	75
		Monsanto	32
		Allied	25
		Hoechst-Celanese	18
	Western Europe	BASF	86
		Bayer	36
		Montedison	32
		Rhône-Poulenc	25
	Japan	Toray	23
		Ube	19
		Unitika	13
		Mitsubishi Kasei	10

Commonly Used Abbreviations for Industrial Polymers

AAS	Acrylate-acrylonitrile-styrene terpolymer
ABS	Acrylonitrile-butadiene-styrene terpolymer
ACS	Acrylonitrile-chlorinated polyethylene-styrene terpolymer
ASA	Acrylate-styrene-acrylonitrile terpolymer
BR	Butadiene rubber
CPE	Chlorinated polyethylene
CPVC	Chlorinated polyvinyl chloride
EAA	Ethylene-acrylic acid copolymer
EEA	Ethylene-ethyl acrylate copolymer
EMA	Ethylene-methacrylate copolymer
EPDM	Ethylene-propylene-diene terpolymer
EPM	Ethylene-propylene copolymer
ECTFE	Ethylene-chlorotrifluoroethylene copolymer
EVA	Ethylene-vinyl acetate copolymer
EVOH	Ethylene-vinyl alcohol copolymer
FEP	Fluorinated ethylene-propylene copolymer
HDPE	High density polyethylene
HIPS	High impact polystyrene
HMW-HDPE	High molecualr weight, high density polyethylene
LDPE	Low density polyethylene
LLDPE	Linear low density polyethylene
MABS	Methyl methacrylate-acrylonitrile-butadiene-styrene polymer
MBS	Methyl methacrylate-butadiene-styrene terpolymer
MF	Melamine-formaldehyde resin
MMA/EA	Methyl methacrylate-ethyl acrylate copolymer
NBR	Acrylonitrile-butadiene rubber (nitrile rubber)
PAN	Polyacrylonitrile
PB	Polybutadiene
PBT	Polybutylene terephthalate
PC	Polycarbonate
PCTFE	Polychlorotrifluoroethylene
PE	Polyethylen
PEG	Polyethylenen glycol
PET	Polyethylene terephthalate
PF	Phenol-formaldehyde resin
PIB	Polyisobutylene
PIR	Polyisocyanurate foam

Commonly Used Abbreviations for Industrial Polymers

PMMA	Polymetyl methacrylate
PP	Polypropylene
PPG	Polypropylene glycol
PPO	Polyphenylene oxide
PPS	Polyphenylene sulfide
PS	Polystyrene
PTFE	Polytetrafluoroethylene
PTMG	Polytetramethylene glycol
PU	Polyurethane
PVA	Polyvinyl acetate
PVAL	Polyvinyl alcohol
PVB	Polyvinyl butyral
PVC	Polyvinyl chloride
PVDF	Polyvinylidene fluoride
PVF	Polyvinyl fluoride, polyvinyl formal
SAN	Styrene-acrylonitrile copolymer
SBR	Styrene-butadiene rubber
TPU	Thermoplastic polyurethane
UF	Urea-formaldehyde resin
UHMW-HDPE	Ultrahigh molecular weight, high density polyethylene
ULDPE	Ultra-low-density polyethylene

Index

AABB-type polycondensation 133
AB polyamides 134
ABS/nylon alloys 84
ABS/PC alloys 84
ABS/polysulfone alloys 84
ABS/polyurethane alloys 84
ABS/PVC alloys 84
ABS/SAN alloys 84
Acetylene 26
Acrylic adhesives 90
Acrylic fibers 89
Acrylic polymers 14, 88
Acrylonitrile-butadiene copolymers 76
Acrylonitrile-butadiene-styrene terpolymers (ABS) 14, 70
Acrylonitrile-chlorinated polyethylene-styrene terpolymer (ACS) 54
Addition polymerization 17
Aliphatic-aromatic polyamides 135
Alkyd resins 127
Amorphous polymers 20
Anion-exchange resins 168
Anionic chain polymerization 48
Anionic polymerization 43
Architectural sealants 108
Aromatic polyamides 135
Aromatic Polyesters 127, 156
Aromatic polymers 150
ASA/PC blends 84
Atactic polymers 47

Benzene 25
Bicomponent fibers 120, 133
Biopolymers 171
Bisglycidyl ethers 113
Bisglycol methacrylates 90
Black orlon 150, 151
Blends 80, 85
Block copolymers 50, 68, 69
Blow molding 30, 31, 32
BTX (benzene, toluene, xylene) fraction 24
Bulk polymerization 27
Butadiene Rubber (BR) 63
Butyl Rubber 77

Calendering 30, 32
Carpet underlay 108
Catalloy 81
Cation-exchange resin 168
Cationic chain polymerization 50
Cationic polymerization 43
Celluloid 13, 125
Cellulose 13, 22
Cellulose acetate 13, 125
Cellulose butyrate 125
Cellulose copolymers 125
Cellulose esters 124
Cellulose nitrate 125
Cellulose triacetate 125
Ceramics 157
Chain flexibility 20
Chain-growth mechanism 17
Chain polymerization 43
Chain transfer agents 73
Chlorinated polyethylenes 18, 19
Coal 26
Commercial isocyanates 101

Compression molding 30, 33
Condensation polymerizations 17
Copolymers of isoprene 65
Copolymers of polyarylates 128
Crystal PS 61
Crystalline polymers 20
Crystallinity 52
Cyanoacrylates 90

Degradable plastics 37

Elastomers 20
Electroconductive polymers 163
Electrodeposition coatings 113
Electrophotography 162
Emulsion polymerization 28, 44
Engineering plastics 20, 110, 127
Engineering thermoplastics 116, 154
EPDM/PP alloys 81
Epichlorohydrin polymers 87
Epoxies 14
Epoxy Resins 112
Ethyl acrylate-methyl methacrylate copolymers (MMA/EA) 83
Ethyl cellulose 125
Ethylene 24
Ethylene copolymers 53, 54
Ethylene glycol dimethacrylates 90
Ethylene-methacrylic acid copolymers (ionomers) 71
Ethylene-propylene copolymers (EPM) 76
Ethylene-propylene elastomers 76
Ethylene-propylene-diene terpolymers (EPDM) 71, 76
Expandable polystyrene (EPS) 61, 62
Extrusion 30, 32

Fiber-reinforced plastics (FRPs) 136
Fillers 29
Film extruders 32
Flexible polyurethane foams 102, 103
Fluoroelastomers 79
Fluoropolymer alloys 87
Formica 144
Free radical 43

Gel coat resins 127
Genetic engineering 172
Glass transition 18
Graft copolymers 68
Graphite HM 150
Graphite HT 150

HDPE/PIB alloys 82
Head-to-head polymerization 45
Head-to-tail polymerization 45
Heterocyclic polymers 156
High density 51
High impact polystyrene (HIPS) 82
High molecular weight (HMW) HDPE 52
Hivalloy 82
Hollow-fiber membranes 167
Hydrogel polymers 93
Hypalon 54
Hytrel 124

Impact polystyrene (IPS, HIPS) 61
Incineration 37
Injection molding 30, 31

Index

Inorganic polymers 157
Interfacial polymerization 28
Interpenetrating networks (IPNs) 80
Ion-exchange resins 167
Ionomers 54, 82, 85
Isophthalic acid 24
Isotactic polyolefins 21
Isotactic polypropylene 47, 59

Kevlar 136
Kevlar fibers 136
Kraton 82, 87

Ladder polymers 150, 151
Lignin 22
Linear low density polyethylene (LLDPE) 51
Liquid crystal polymers (LCPs) 156
Living polymers 50
LLDPE blends 82

Macrocyclic ureas as masked diisocyanates 107
Melamine-formaldehyde resins (MF) 14, 144
Melt viscosity 17
Merrifield resins 13, 170, 172
Metallocene catalysts 52
Methane 25
Methyl methacrylate-acrylonitrile-butadiene-styrene (MABS) 83
Methyl methacrylate-butadiene-styrene (MBS) 83
Millable polyurethane elastomers 105
Modacryil fibers 89
Modified PPO 86, 116
Moisture-cured polyurethane coatings 107
Molecular weight 17

Natural gas 23, 25
Neoprene 14, 66
Nitrile rubber (NBR) 76
Nomex fibers 135
Nonwoven fabrics 121
Noryl 86
Novolacs 142
Nylon-4,6 133
Nylon-6 132
Nylon-6,6 132
Nylon-11 133, 165
Nylon-12 134
Nylon-alloys 85
Nylon/ionomer blends 85
Nylon-polypropylene alloys 85
Nylon/PTFE alloys 85

Oligomers 17
Oligonucleotides 172
Organometallic chain polymers 157
Organometallic initiation 43

PBT/PET blends 86
PBT/styrene-maleic anhydride copolymer blends 86
PC/PBT alloys 85
PC/PET alloys 86
PC/styrene-maleic anhydride copolymers 86
Peptides 13
Peptide synthesizers 172
Permanent-press fabrics 121
Pervaporation 167
Petroleum 23
PHD polyethers 112
Phenol-formaldehyde resins (PF) 13, 141
Phenolic foams 141
Photocrosslinking process 165
Photofabrication 166

Photopolymer printing plates 165
Photopolymerization 165
Phthalic anhydride 24
Piezoelectric polymer 165
Poly(1-butene) 60
Poly(2-ethylhexyl acrylate) 91
Poly(3-alkylthiophene) 163
Poly(3-hydroxybutyrate) 125
Poly(4-methylpentene) 60
Poly(ethylene-co-acrylic acid) EAA 92
Poly(ethylene-co-ethyl acrylate) EEA 92
Poly(ethylene-co-vinyl acetate) EVA 96
Poly(sulfur nitride) 161
Poly(p-phenylene) 150
Poly(p-xylylene) 150
Polyacetal 14
Polyacetylene 163
Polyacrylamide 94
Polyacrylates 91
Polyamides 14
Polyamide imides 138
Polyamidoamines (PAMAM) 171
Polyaniline 164
Polyarylamides 15
Polyarylates 15, 127
Polyarylketones 15, 152
Polyarylsulfones 15
Polybenzimidazole 15, 156
Polyboron nitride 157
Polybutadiene 48, 63
Polybutylene terephthalate (PBT) 15, 123
Polycarbonate (PC) 14, 128
Polycarborane-siloxanes 160, 161
Polychloroprene 66
Polychlorotrifluoroethylene (PCTFE) 57
Polycyclodisilazanes 159
Polydihydroxymethylcyclohexyl terephthalate 124
Polydimethylsiloxane 159
Polyesters 14
Polyester alloys 86
Polyether ether ketone (PEEK) 151
Polyether ketone (PEK) 152
Polyether polyols 111
Polyether sulfone 154
Polyethyl acrylate 91
Polyethylene (PE) 14, 45, 50, 51
Polyethylene terephthalate 14, 122
Polyethylene terephthalate fibers 120
Polyfluorosilicones 79
Polyimides 15, 139
Polyisoprene 65
Polyisoprene elastomers 66
Polyisothianaphthene 164
Polyketones 54
Polymer alloys 80
Polymer polyols 103, 111
Polymeric hydrogenation catalysts 170
Polymeric reagents 169
Polymethyl acrylate 91
Polymethyl methacrylate (PMMA) 93
Polynorbornene 45
Polynucleotides 172
Polyolefins 41, 42, 44
Polyolefin alloys 81
Polyoxymethylene (polyacetal) 109
Polypentene-2 48
Polypeptides 173
Polyperfluorocarboxylic acid 169
Polyphenylene 163

Polyphenylene oxide (PPO) 116
Polyphenylene oxide alloys 86
Polyphenylene sulfide (PPS) 15, 152, 164
Polyphenylene vinylene 163
Polyphosphazenes 160
Polypropylene (PP) 14, 59
Polypyrrole 163
Polyquinoxaline 151
Polysaccharides 172
Polysilanes 159
Polystyrene (PS) 13, 61
Polysulfones 153
Polytetrafluoroethylene (PTFE) 45, 58
Polytetrafluoroethylene-co-perfluoromethyl-vinyl ether 79
Polytetrafluoroethylene-co-propylene 79
Polytetramethylene glycol (PTMG) 112
Polythiazyl 161, 164
Polythienyl vinylene 164
Polyurea RIM systems 105
Polyurethanes 14
Polyurethane coatings 106
Polyurethane elastomers 105
Polyurethane foams 29
Polyurethane powder coatings 107
Polyurethane spandex fiber 106
Polyvinyl acetals 98
Polyvinyl acetate (PVA) 14, 96
Polyvinyl alcohol (PVAL) 28, 97
Polyvinyl butyral 98
Polyvinyl chloride (PVC) 13, 55
Polyvinyl chloride alloys 83
Polyvinyl cinnamates 165
Polyvinyl ethers 98
Polyvinyl fluoride (PVF) 54
Polyvinyl formal 98
Polyvinyl pyridine 62
Polyvinylcarbazole 99, 162
Polyvinylidene chloride (PDC) 57
Polyvinylidene fluoride (PVDF) 56, 164
Polyvinylidene fluoride-co-chloro-trifluoroethylene 79
Polyvinylidene fluoride-co-hexafluoro-propylene 79
Polyvinylidene fluoride-co-hexafluoro-propylene-co-tetrafluoroethylene 79
Polyvinylidene fluoride-co-tetra-fluoroethylene-co-perfluoromethyl vinyl ether 79
Polyvinyl pyrrolidone (PVP) 99
Polyarylates 128
Powder coatings 115
Propylene 24
PTFE alloys 87
PVC plastisols 56
PVC/CPE alloy 83
PVC/EPDM graft copolymers 83
PVC/MMA blends 83
PVC/NBR blends 83
Pyrolysis 37

Qiana fibers 134

Random copolymers 68
Reaction injection molding (RIM) 33, 102
Reaction injection molding (RIM) process 30
Recombinant DNA technology 173
Recycling 35, 36, 37
Reinforced polyesters 126
Resoles 141
Rigid polyurethane foam 102, 104
RIM polyurethane systems 105

RIM products 105
Ring-opening polymerization 14, 45
Room temperature vulcanizing rubbers (RTV) 158
RRIM (glass-reinforced RIM) 105
Rubbery state 18

SAN copolymers 70
SAN/EPDM blends 84
Santoprene 81
Semirigid polyurethane foam 103
Sheet molding compound 34, 126
Silicones 158
Silicone elastomers 159
Silicone fluids 159
Spheripol process 59
Spinning methods 34
Star block copolymers 68
Starburst dendrimers 171
Starch 172
Step-growth-process 17
Stereoregular polymers 46
Stereospecific polymerization 14
Structural adhesives 108, 112
Styrene-acrylonitrile copolymers (SAN) 69
Styrene-butadiene block copolymers 74
Styrene-butadiene rubber (SBR) 73
Styrofoam 62
Super tough nylons 85
Surlyn copolymers 71
Suspension polymerization 27, 44
Syndiotactic polypropylene 47
Synthesis gas 25
Synthetic fibers 34, 120

Tacticity 47
Temperature 18
Terephthalic acid 24
Tetrafluoroethylene copolymers 78
Thermal depolymerization 37
Thermoformable flexible foams 71
Thermoplastic 17
Thermoplastic olefin elastomers (TPOs) 77, 81
Thermoplastics polyurethane elastomers (TPUs) 105
Thermoset 17
Trichlorofluoroethylene copolymer with ethylene (ECTFE) 57
Tyrin 54

Udel 154
Ultem 140
Ultra-low-density polyethylene (ULDPE) 51
Ultrahigh molecular weight (UHMW) HDPE 52
Unsaturated polyesters 126
Urea-formaldehyde resins (UF) 13, 143
Urethane-modified isocyanurate foams 104

Vectra 156
Vesicular images 166
Vinyl pyridine-butadiene-styrene terpolymers 62
Vinylon fibers 98

Water-based polyurethane coatings 108

Xydar 156

Ziegler-Natta catalysts 43, 47, 51

WEST GEORGIA TECH LIBRARY
FORT DRIVE
LAGRANGE, GA 30240